LAND, WEATHER, SEASONS, INSECTS: AN ARCHETYPAL VIEW

Also by Dennis Merritt

Jung and Ecopsychology
The Dairy Farmer's Guide to the Universe Volume I
ISBN 978-1-926715-42-1

The Cry of Merlin: Jung, the Prototypical Ecopsychologist
The Dairy Farmer's Guide to the Universe Volume II
ISBN 978-1-926715-43-8

Hermes, Ecopsychology, and Complexity Theory
The Dairy Farmer's Guide to the Universe Volume III
ISBN 978-1-926715-44-5

LAND, WEATHER, SEASONS, INSECTS: AN ARCHETYPAL VIEW

THE DAIRY FARMER'S GUIDE TO THE UNIVERSE VOLUME IV

DENNIS L. MERRITT, PH.D.

Land, Weather, Seasons, Insects: An Archetypal View
The Dairy Farmer's Guide to the Universe Volume 4

First Edition
ISBN 978-1-926715-45-2 Paperback

Published simultaneously in Canada, the United Kingdom, and the United States of America by Fisher King Press. For information on obtaining permission for use of material from this work, submit a written request to: permissions@fisherkingpress.com

Fisher King Press
PO Box 222321
Carmel, CA 93922
www.fisherkingpress.com
info@fisherkingpress.com
+1-831-238-7799

CONTENTS

The four volumes of *The Dairy Farmer's Guide to the Universe* offer a comprehensive presentation of Jungian ecopsychology. Volume 1, *Jung and Ecopsychology*, examines the evolution of the Western dysfunctional relationship with the environment, explores the theoretical framework and concepts of Jungian ecopsychology, and describes how it could be applied to psychotherapy, our educational system, and our relationship with indigenous peoples. Volume 2, *The Cry of Merlin: Jung, the Prototypical Ecopsychologist*, reveals how an individual's biography can be treated in an ecopsychological manner and articulates how Jung's life experiences make him the prototypical ecopsychologist. Volume 3, *Hermes, Ecopsychology, and Complexity Theory*, provides an archetypal, mythological and symbolic foundation for Jungian ecopsychology. Volume 4, *Land, Weather, Seasons, Insects: An Archetypal View* describes how a deep, soulful connection can be made with these elements through a Jungian ecopsychological approach. This involves the use of science, myths, symbols, dreams, Native American spirituality, imaginal psychology and the *I Ching*. Together, these volumes provide what I hope will be a useful handbook for psychologists and environmentalists seeking to imagine and enact a healthier relationship with their psyches and the world of which they are a part.

My thanks to Craig Werner for his comprehensive and sensitive editorial work, and to Tom Lane for his constructive comments.

To Werner Loher and James Hillman, two real and imaginal mentors, and to Gene Defoliart, a good old boy in a good way, from Arkansas.

The environmental problem has religious as well as scientific dimensions...As scientists, many of us have had a profound experience of awe and reverence before the universe. We understand that what is regarded as sacred is more likely to be treated with care and respect. Our planetary home should be so regarded. Efforts to safeguard and cherish the environment need to be infused with a vision of the sacred. At the same time, a much wider and deeper understanding of science and technology is needed. If we do not understand the problem it is unlikely we will be able to fix it. Thus there is a vital role for both science and religion.

—Carl Sagan, 1992

CHAPTER 1

An Archetypal View of the Midwest Environment

Everyone has to feel rooted, has to have a foundation for their life. The roots can be in family, friends, home or community. For many, deep roots extend into the land and the region they live in. Two important ecopsychological premises are that we are capable of far deeper connections to the land and those who feel connected to the land are naturally interested in protecting and preserving it. The Midwest is the area I feel rooted in. When I grew up on a small Wisconsin farm in the 1950's and 1960's there was an ecopsychological balance between the fecundity of the land—the archetype of the Great Mother in her nourishing form—and the archetype of the farmer in intimate relationship with her.

What is so compelling about a region considered by many to be flyover country? A numinous dream of a typical meadow in Wisconsin enticed me to look at the Midwest more closely, feel more deeply for its essence, and appreciate its many dimensions. I use a variety of approaches to bring to consciousness what is special about the Midwest environment, approaches that can be used anywhere to deepen a connection with the land and establish a sense of place. This is not insignificant in the mobile, rootless society that is America.

To sense the spirit of a place, I analyze an environment as if it were a dream. Like a dream, I consider every element in the environment for its potential depth and interior nature. I look for metaphors contained in the image, its "as if" dimension. What scientific facts will deepen an understanding and appreciation of the image? What are its symbolic dimensions? How does it appear in myths and stories? The result is that one can "see into" the environment and experience its essence. With this approach I will consider the Midwest's location, glacial history, topography, water systems, dominant plants and animals, and agriculture. Weather and seasons will be examined in the next chapter, "Seasons of the Soul."

I. Location

The broadest and most general aspect of the Midwest is its location. Physically it is the center of America, its heartland. As in working with dreams, listen for overtones and undertones of words; sense the metaphors and the deeper psychological meanings. We're talking about America's Heartland, i.e., close to the heart, adored, a core feeling, including how the country feels about itself in its heart. The Midwest is in the center of the country, implying central, correct, basic and essential. Think of the opposites of words to get a better sense of their meaning. Opposite of central would be fringe, as in "fringe elements"; border, as in borderline sanity; far out, less important, less essential. We are in the middle of the land mass called North America, Turtle Island to many Native Americans; continental, massive, not easily moved, immovable, stable. By contrast, from the perspective of many Midwesterners, California is "the land of flakes and quakes."

Midwest conjures up a psychological locale between the established, the city, the older and stogy East, and the wild, uncultivated, younger, brash, more irresponsible West. (Shortridge 1989, p. 8) The Midwest is associated primarily with a massive agricultural base that is cultivated, prosperous and stable. It is part of the myth of the developmental stages of the life of a country—youth, middle age and old age going from West to Midwest to East.

II. Midwest Topography

We'll start on the environment of the Midwest by going from the ground up, working on the premise that every landscape has a soul. The most general fact about the Midwest's physical environment is its basic topography. It is like a gigantic shallow trough slanting to the middle and south (except along it northeastern fringe) stretching from the ancient Appalachians in the east to the newer Rocky Mountains in the west. (1) This is the twelve state area going south from North Dakota to Kansas and east to Ohio. Much of the Midwest had been part of an ancient inland sea located on the equator half a billion years ago. Infinite numbers of tiny creatures laid down a limestone bedrock. The northernmost Midwest is part of the Canadian Shield, a bedrock of ancient hard rocks like granite and covered with shallow soils in northern Minnesota, northern Wisconsin and the western portion of Michigan's upper peninsula.

The sharpest metaphoric contrast to the level, almost flat topography of the Midwest is mountains. The basic landscape of the Midwest is not one of highs, peak experiences, elevations and depressions, or ruggedness. Compared to the insular, inclosing effect of mountain valleys, the flatter landscape is something more open, expansive and unprotected. Mountains and mountaintops are generally associated with spirit. It helps in developing metaphors to "try" on several images of the same theme. Imagining the type of person living in the mountains leads one to think of a rugged, adventuresome individual, taking more risks by driving mountain roads and mountain climbing. We might see him or her as a survivor of harsh, challenging environments; a rough-and-ready person, a Marlboro Man, probably riding a horse or driving a beat-up 4-wheel-drive vehicle.

Contrast this with someone living on a rather flat landscape. We might imagine someone more down to earth, earthy, not extreme in highs and lows; someone with flatter affect, not radical or extremist, but moderate, middle of the road, maybe bland, maybe someone more materialistic. Each soul will feel more at home and develop in a particular landscape that suits it, needing occasional visits to contrasting environments to add other dimensions.

III. Glacial History

One cannot appreciate the Midwest without understanding its glacial history. Every Midwestern state has been profoundly affected by the ice giants that rolled through here in recurrent waves eons ago. An analyst friend dreamt she should live where the glacier had been. To sense the depth of this dream we must re-imagine glacier country. We ask along with her, "What is the soul of glacier country? How does a glacial landscape impact the psyche?"

Learning about glaciers and recognizing the evidence of their past helps one resonate with the soul of glacier country. The signs are everywhere if one has the eye to see. Gravel pits, those convenient entrances into the underworld, reveal a thick layer of soil called glacial till created by that giant land scraper, the glacier. The soil is a mixture of material the glaciers scoured off the landscape in their marches down from Canada. Scattered throughout glacial till are rocks rounded by tumbling in the bowels of the ice-giants. Most are scarred by straight lines scratched across them by other rocks. Their mineral content and composition belie an ancestry different from the bedrock layer of limestone indigenous to the landscape's foundation. These unpretentious gravel pits reveal a history of vast changes from long ago.

But not that long ago. It has been only 11,000 years since the last glacier melted and ran off the upper Midwest landscape. The landforms glaciers created serve as reminders of the end of the age of mastodon hunters and the beginning of agriculture and civilization. Geologists tell us there have been several glacial periods within the current Ice Age, the Quaternary, that began about 2-1/2 million years ago. The southern-most advance of the glacial mass in the respective periods reached Kansas, Nebraska, Illinois and Wisconsin. The last glacial period, the Wisconsin, ended about 11,000 years ago, having begun 110,000 years ago. Wisconsin has risen 160 feet since getting all that ice off its back and is still rising 1/2 inch a year: it was that depressed by the glacier.

The soul of the Midwest is in many ways indebted to the past lives of glaciers that have come and gone over millions of years for reasons unknown. Much of the Midwest had been part of a vast inland sea hundreds of millions of years ago. The relatively flat sea bottom left behind from the drainage of the ancient ocean was further leveled by the onslaughts of water in solid moving form. But it is not fair to say that glaciers only flatten a topography; they do provide some relief. Moraines are hills constructed from debris scoured off the land and dumped at the glacier's melting end point or the squeeze line between the lobes of a glacier. These hill formations have an irregular surface and are imaginatively arranged: a jazz landscape. You know when you're in moraine country when you start seeing a lot of gravel pits—open pit mining of glacial debris. Moraines often contain interesting features such as kettles: big, round pits formed when huge chunks of debris-buried ice took a thousand years to melt, creating a glacial sinkhole. (Fig. 1)

Erratics are huge rocks carried great distances, sometimes hundreds of miles in or on the glacier. They get deposited on the land in an erratic pattern when the glacier melts, left sitting there atop the soil. These lost souls fascinate the psyche, many of the larger ones being sacred to the Native Americans in this region.

Ice a mile or more thick spread out over half a continent has interesting things happening within it. The hidden activities of glaciers are revealed upon melting, like remembering a dream after we wake from slumber. Rivers form inside a glacier whose riverbeds gradually fill with debris. When the glacier melts, these riverbeds of debris are dumped on the land, creating snake-like hills called eskers crawling across the landscape. Cone-shaped hills called kames form when debris carried by meltwater washes through holes in the ice and settles into an ice cavern below. It forms an inverted V, like sand settling on the bottom of an hourglass, which gets dropped on the landscape when the glacier disappears.

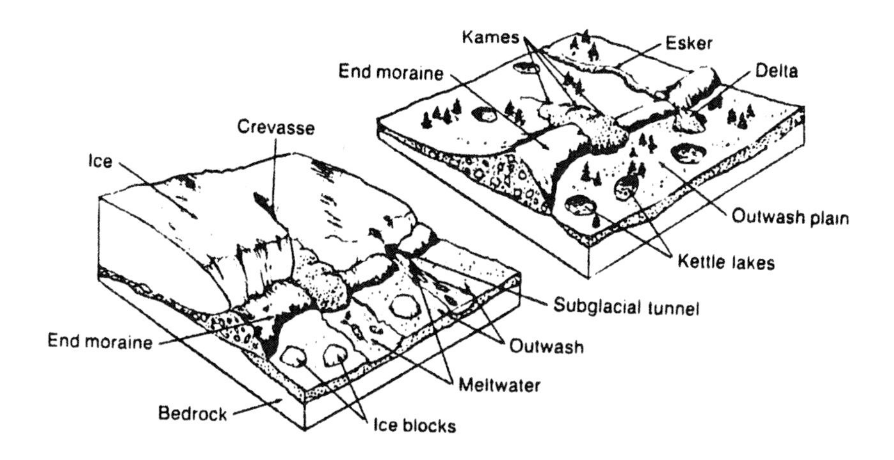

Figure 1. Glacial and post-glacial landscapes. (Sullivan 1984, p. 253)

My favorite glacial landscape is in drumlin country. Scientific imaginations still haven't figured out how some of the hills were formed. It's believed they were built up in layers beneath the moving glaciers. They occur in groups, sometimes in the hundreds, behind the terminal moraine. From above they look like long, narrow, semi-flattened teardrops best imagined as an entire fleet of gigantic battleships lined up in the same direction. The tapered end of the teardrop points to the direction the ice was moving, like the V of water that forms behind an exposed rock in a river. You don't have to get too high in an airplane to see geological history being pointed out by clusters of these massive markers. The I-94 freeway between Madison and Milwaukee, Wisconsin should be named "the drumlin special"—it cuts straight through some of the finest drumlin scenery in the world.

When I look out over a drumlin countryside I see a vast sea of gigantic land waves frozen in earth time. These paradoxical waves formed under water in its solid form. I get an almost mythical sense that a giant hand played in a sandbox we now call Wisconsin to create these formations.

I am aware of immense cycles of time in drumlin country. These hills are a reminder that in this place, just over 11,000 years ago, was a pile of ice several thousands of feet thick. It is also a reminder that we don't know the mysteries of earth's warming and cooling spells that extend over eons of time. Every spring when I see the glacial hills greening, I'm thankful for the gift of warmth returning life to glacier country. Does climate change sound the death knell for glaciers?

The varying glacial topographies generate subtle but significant effects on the earth-sky relationship. Glacial hills, often decorated by their crowns of trees, are just high enough to intrude into the sky horizon. The sky becomes a backdrop leaving the earth to have the prominent effect on the psyche. It doesn't take much to lose the sky as the dominant element. The contrast is seen in the flatter Illinois prairie country which presents a Big Sky effect on a state-wide basis.

The low elevations of the hills in the Midwest give the land a very human dimension. Contrast this with the Rockies, where one is overwhelmed by their awesome size and grandeur. Glacial hills gently surround and comfort the psyche. The rather flat or gently hilly topography allows it to sustain non-permanent vegetative cover. This means the landform is suitable for farming because erosion can be limited, the consequence being that cultivated plant life is introduced to the psyche.

A secondary yet important agricultural effect created by the glacier is not noticeable unless studied scientifically. Glaciers tilled the soils, generating a good physical mix of particle sizes and minerals. The result is an adequate physical base for topsoil development and plant growth. One realizes the significance of this glacial mixing when considering the problems for plants trying to grow on sandy soils, or the wastelands created after some rain forests are cut and the clay hardens into a form forever useless for farming.

Understanding how the features in glacial landscapes were created, the vast time frames involved, the information buried beneath our feet; this is knowledge generated by science. This is science with soul, information that connects us to things and adds an awesome dimension to the lives of those who know, who see.

IV. The Water Environment

There is an important interactive effect of glaciers to consider in the Midwest—they gather water. Glaciers ruined the arboreal (tree-like) drainage system of little streams leading to larger streams feeding increasingly larger rivers. Glaciers cut across these watery limbs, forcing the water to congregate into myriads of lakes, ponds, potholes, kettles and marshes. The Great Lakes lie in giant beds carved out by the glaciers.

The particular size and shape of the watery container have specific effects on the psyche. Lakes in contrast to oceans are bounded waters, containers of human-dimensional sizes. The human mind can envelop a lake, can wrap itself around it. Oceans are immense, impersonal; they generate feelings of cosmic forces, of extraterrestrial influences from the moon and archetypal ebbs and flows. Their salty water is not "fresh" and does not nourish human life.

Water is the second vital element in the Midwest. It's an important resource here. Californians had to build a 400-mile aqueduct to carry water from northern to southern California. Midwest water is from rain, not from irrigation or pumps except on the western fringe. We get adequate rainfall that can support abundant plant life, about 30 inches a year well distributed. The Midwest has the necessary water of life—fresh, drinkable water.

Midwesterners need help to appreciate hot, muggy summer days. They have to recognize that such weather is the invisible pipeline to a prime water source—the Gulf of Mexico. We have the occasional drought in Wisconsin. One gorgeous summer day follows another with moderate temperatures, sunshine and low humidity—and plant life suffers. Much of our water of life is rung out of muggy summer days by thunderstorms. One is more appreciative of muggy days following a terrible drought.

Our abundance of rainfall fills up the innumerable variety and shapes of water beds created by the glaciers, including that massive gathering of waters called the Great Lakes. Here we have the water of life on a grand scale, containing almost one-fifth of the planet's fresh water. The

Great Lakes States have a different type of continental environment, with moderated weather within miles of the lakes. A Windy City is created by the cooler lake air deflecting transient air masses around the southern tip of Lake Michigan. It is a maritime psyche of mammoth lake storms—some of the worst in the world— huge fish, ocean-going ships, and port cities with all that such cities imply.

We have to bring to bear all our senses to imagine water in the Midwest. Consider being in a boat on a hot day or swimming—fluidity, dissolving, splashiness, playfulness, invigorating coolness. We are buoyed up, supported. Midwest waters contain an abundance of plant and algal life, they're teeming with life; not like crystal-clear mountain lakes, but murky and rich with organic watery smells.

Some of the bogs, potholes and ponds are actually kettles—those glacial sinkholes formed when huge chunks of buried ice finally melted. These waterholes and marshlands are rampant with life and are important for birds and purifying water. They ooze rich, mucky smells from ages of decaying vegetation. Smells are hard to describe. The sense of smell is direct, strong, immediate, particular, and can reach us from unseen sources. It is a primal, animal sense often feared by over-spiritual types, especially the lusty smells of sexual and earthy activity. The dead in Hades orient by smell and so do salmon in finding their river homes after ocean or Great Lakes journeys.

There are a wealth of rivers here despite the effects of the glaciers. A woman friend who lives by the Mississippi River loves sitting next to the river—it has a wonderful effect on her psyche. Where does this effect come from? The movement of water is a prime metaphor about life for the Taoists and a river is an ideal metaphor for the flow of life. Rivers present the paradox of being in place yet ever moving, an analogy of the psyche being in our physical body but ever changing.

V. Forests and Farms

Elemental forces and forms have combined in the right proportions in the Midwest to help create its most distinctive feature—agriculture, a cultivated landscape. Glacial land is symbiotically joined with abundant rainfall and the heat from our local star that produces warm summers and a long, frost-free growing season. Severe winters rid the realm of many insect pests. American scientists, farmers and agribusiness working with the natural blessings that are the Midwest have produced a remarkable phenomenon—a cornucopia of food, the food for human life. Animals are fattened from the harvest of the extremely fertile lands of the Corn Belt (for corn and soybean production) which includes the whole state of Iowa, most of Illinois and Indiana, substantial portions of Minnesota and Nebraska, and good chunks of the other Midwestern states. The bread of life comes from the Wheat Belt that abuts the western end of the Corn Belt, and dairying in Wisconsin and Minnesota crowns its north central portion. The Midwest contains the most vast, fertile and prolific agricultural area in the world. There's nothing that approaches it in Canada, Mexico, Africa, Europe (including the Ukraine), Asia or Australia. A challenge to the Midwest supremacy in farmland is arising in the soybean and cattle ranching acreage created in the wake of the slaughter of the Amazon rain forest.

Not all of nature in the Midwest has been worked over to nourish humankind. There are some wild, uncultivated oases remaining in this reconstructed landscape. These are the great north woods of Minnesota, Wisconsin and Upper Michigan. The shorter growing season in the far north and the poor soils over the Canadian Shield made this land unattractive for farming, so the land was returned to forests after Paul Bunyan left and farms failed. The great Giants of the upper Midwest forests are gone, their descendents marketed out or ending up in fireplaces before they reach the elder stage. A few of them remain, even in cities, standing as mute sentinels to haunt us with a memory of a past grandeur.

To sense the impact of forests on the psyche, we have to imagine what it is like to be in a forest—a world of shadows, filtered light, wild and large animals lurking and hidden; of seasonal cycles of forest wildflowers, berries and birds. The glaciers left beds for the innumerable lakes and bogs in these forests. Bog smells are distinct. Spruce trees struggle to survive in them. If these trees were people, they would be charter members of Adult Child groups.

Groves of trees and woodlots in farm country are often indicators of rivers or indirect signs of glacial remnants. Farm woodlots are relegated to the least arable land—too hilly and/or with stony or poor soil—attributes of glacial moraines or steep drumlins. A loosely continuous band of woodlots atop hills often mark a line of glacial moraines, those debris dumps that formed at the end or between lobes of long extinct glaciers.

It was in our farm woodlots and "useless marshland" that I forged my deepest connection to nature. I spent many hours wandering these marginal areas with my faithful dog, each of us curious and excited by the "wild." A river ran through the home farm, adding another dimension to our domain—different sounds, smells, insects, birds, plants, animals, fish and snapping turtles.

We've seen how the combination of climate, soil, topography and water give the Midwest its most distinctive feature—farmland. To extend our imagination deeper into farmland, we must first consider whether farm animals are present. One gets a very different feel from the landscape depending on whether or not the prominent Midwest animals are around—the cow or the pig. Cows especially will animate a rural environment, giving it a more pastoral setting.

Let's begin by thinking about the cow. Thinking cow is not to think too much, for cows aren't the most intelligent of beings. But they're bright enough for what they have to do—eat and make milk. These ungainly looking animals have big stomachs (four in fact), a big udder, big nose, big eyes and big ears. Our bovine wonders produce 20% of America's agricultural wealth. Over 30% of the actual food value consumed by Americans comes from dairy products and 4 million people are directly or indirectly employed making them. (Rath 1987, p. ix)

To think cow is to think pasture, meadows, pastoral; grazing, grazing in the grass; feeding, feeding those enormous stomachs. A cow has to eat a lot to produce all that milk. In the mind's eye one sees a sunny day with a herd of cows moving slowly across a hilly field, heads to the ground, packing it away. Notice, we thought of a *herd* of cows—cows are not solitary animals. They are "groupies" with cow pecking orders on a much grander scale than chicken pecking orders. They don't peck; they buck each other which is why they're dehorned in infancy. Cows certainly aren't fleet footed, being by nature bulky and slow moving. Rural humor is a herd of cows being chased by deer flies. Only then will you see cows gallop as they race through the pastures with tails aloft, teats flying every which way, milk shakes in the making.

Thinking cow slows you down, grazes you out, makes you think about eating, or more likely—ruminating. Cows are archetypal ruminants; they belong to a group of animals, such as goats and sheep, that ruminate. A cow stuffs the first of her four stomachs with grass or alfalfa, gulping it down after grasping it with her huge, wet, coarse sandpaper tongue and biting it off against a toothless upper front gum.

Cows are smarter than we think: who wants to be out in a hot summer sun on a muggy day moving a blimp of a body around on slender legs? You could be resting under a nice shade tree while chewing your food properly and thoroughly after regurgitating all that fast food from stomach number one.

Notice we're talking female here. Bulls have been removed to stud farms where technology has decimated their sex lives through the wonders of fake mounts, frozen semen and artificial insemination.

The cow's product, milk, has been advertised as nature's most nearly perfect food: pure, white, wholesome; nutritious from babies through adulthood. Milk is high in calcium needed to build boney structures, rich in proteins to build muscle and flesh, and high in vitamins A, B, C and D.

Given the importance of the cow to humans, it's no wonder she's been highly regarded by the collective unconscious of psyches throughout the world. The ancient Egyptians invoked the feminine side of Nun, the primordial waters, as "the Cow, the Ancient, who bore the sun and set the seeds of gods and men." (Brugsch 1885, p. 114 f. quoted in CW 5, ¶ 358) The goddess Naunet personified the chaotic waters of the beginning, the birth-giving primary substance, and from "Naunet sprang Nut, the sky-goddess, who is represented with a starry body or as a heavenly cow dotted with stars." (Brugsch 1885, p. 128 f. quoted in CW 5, ¶ 359) The great Egyptian goddess Isis was represented in her good mother form with a cow head and hailed as "The First of the Cows." (Neumann 1970, p. 65)

The cow did not go unnoticed in other cultures. The ancient Greeks saw her as the archetypal Earth Mother who suckled Cretan Zeus (Neumann 1963, p. 76) and the cow was one of the early forms of Aphrodite. (p. 81) The wandering milk-white moon cow was Io to the Mycenaeans and the Cretan Europa: the refined subcontinent we call Europe was named after a cow. (p. 80) We poor Americans are left with a nursery rhyme about a cow jumping over the moon.

For the Hindu, "a radiance first came out of the Creator's face and later split into four parts—The Vedas, The Fire, the Cow and Brahmin." Many Sanskrit hymns relate to the cow. The cow world is seen as the highest and greatest of all worlds and cows are spoken of as "mothers to all beings." (Rath 1987, p. 61) Gandhi was more concerned about cow protection than the removal of foreign rule from India. (p. 62)

That's a pretty rich mythological and religious pedigree for this humble creature. There's a lot of truth in our exclamation, "Holy Cow!"

Where you find cows, you're likely to find your basic image of a farm—Old MacDonald's farm. There is often an assemblage of animals on a dairy farm—chickens, ducks, pigs, maybe some sheep. That's because one needs a variety of crops for cows and these can feed other animals too, plus tending to the other animals is minimal compared to cow care. These crops diversify the plant landscape. You don't have monocultures of corn or soybeans on a dairy farm, but a pleasant, edible (to cows) mixture of things: there's corn for silage (*great* smells), oats for ground oats (additional carbohydrate source), and the basic—alfalfa hay, the wonderplant of cow country and the main supply of protein in a cow's diet. The variety of crops, animals and their waste products, i.e., manure, offer a rich palate of smells to the discriminating nose.

Dairying is *the* most demanding farming. The farmer has to be there 365 days a year milking at 5:30 a.m. and 5:30 p.m. (or some other ungodly 12-hour interval). My parents had one vacation together in their 25 years on the farm. Unless cows are familiar with those who milk them, they get anxious and don't release a hormone that allows the milk to be "dropped." If they don't give their milk, they're more likely to get an infected udder, called mastitis, making the cow sore and irritable. She's likely to kick as you try to stick a huge needle up her teats to drain the bad milk. The antibiotic given to cure the mastitis means the milk has to be dumped. Cows demand one big happy family tending them all the time.

This basic farm scene has changed a lot since I was a kid growing up on our family dairy farm. The average was about 30 cows back then, now corporate farms are 1000 cows or more. Huge metal pole sheds and enormous milking parlors have replaced the small red barns that once dotted dairyland. Cows hardly get to graze in green pastures anymore: chopped food is presented to them in dirty feedlots.

There is a preservation effort to keep the old barns from disappearing entirely from our rural image. The small family farm is rapidly going extinct, even in Wisconsin, the Dairy State, where the average age of farmers is over 50. They are waiting to retire and quit the business with no children wanting to take over this economically depressed occupation. Over 40% of the farm workers are immigrants tending the huge herds.

Before we move on to the world of pigs, let's summarize how the cow strikes the human psyche. We would use words like gentle, docile [as described in hexagram 30 of *The I Ching* (Wilhelm 1967, p. 119)], slightly to moderately humorous, ungainly; grazing in the grass, ruminating, lying in the shade on a hot summer day, pastoral; producers of nature's most nearly perfect food (cholesterol concerns aside), and the Great Mother in her nourishing role. We also associate cows with Old MacDonald's archetypal farm and farming at its most varied and intensive.

Ah yes, the pig. What can one "see" in the pig. "Basic pig" is an animal close to the earth thanks to short stubby legs that seem inadequate to hold it off the ground let alone move it along. This proximity to the earth allows it to keep its head close to the ground, hardly needing a neck. Those small eyes don't need to see much beyond their snouts that make them the champion animal rototillers. We're talking archetypal rooter, able to root things up and feed right out of the earth. Their big ears often seem to act as blinders, limiting the pig's worldview even more. Its big belly practically drags the earth, giving an overall appearance of being an eating machine. Witnessing a pig slopping at the trough is to view the origin of the terms "pigging out" and "making a pig of oneself." That belly also gestates a host of piglets, often 10 to 12 at a crack, and all those hungry mouths at the teat counter present an archetypal image of fecundity and Big Momma nourisher. The poor animal doesn't come equipped with a cooling system, so it is driven to wallowing in muck to keep cool even at moderately warm temperatures. Its lack of fur gives it a fleshy presence with no disguises—what you see is what you eat. This earthy creature is finished off in an unusual manner—a curly little tail for humorous decoration only.

"Seeing" the pig makes it clear that this is an animal associated with the earth, female fertility and nourishment. Gimbutas discovered many figures of the pig as symbols of the Pregnant Goddess in the pre-historic goddess cultures of southeastern Europe going back to 7,000 BCE. (Gimbutas 1989, p. 147) Jungian analyst Eric Neumann also emphasized the association of the pig with the female and the fruitful and receptive womb. (Neumann 1970, p. 87)

The ancient Greek *thesmophoria* ceremony clearly showed the pig as a "uterine animal" associated with the earth. The ceremony symbolically fertilized the earth by throwing suckling pig sacrifices into caves, symbolic of the earth's womb. Their rotten remains were placed on alters with pinecones symbolic of the phalli of the tree-son together with wheat cakes in the shape of male genitals, then mixed with seeds to be used for sowing. This was done in honor of Demeter, the Greek Queen of Corn (wheat) and goddess of agriculture, and her daughter Persephone. (Nilsson 1957, p. 312 in Gimbutas 1989, p. 147) Eleusis, site of the ancient Greek mother-cult Eleusinian mysteries, had the pig as a symbol for its mysteries. (Neumann 1970, p. 85) The ancient Egyptian goddess Isis, like Nut, had the "white sow" as one of her forms. Wherever pork is forbidden and the pig considered unclean, we know it was originally a sacred animal. (p. 86)

There are primitive associations of the pig with female genitals, translated from Greek and Latin "pig." (Neumann 1970, p. 85) Even today we use a negative description of someone's sexual behavior as being "swinish." Aphrodite at Argos (Greece) was associated with pig sacrifices and in her original form of the Great Mother she sent "Aphrodisia mania"—the mind boggles with erotic delight! (p. 86)

A personal association of the pig with the earth and the feminine came to me in a dream I had while training at the Jung Institute in Zurich:

> I'm going to visit an older woman who lives at the San Francisco Bay end of University Avenue in Berkeley. She lives in a small wooden house half sunk into the earth. I enter through a low door where I'm met by a big black pig. I'm a little fearful of it but the woman's demeanor conveys a sense of it being O.K. and not to worry about the pig.

University Avenue runs from the San Francisco Bay up to the Berkeley campus set at the base of, and partly in, the hills. The dream locale is at the opposite end of the road from the University, i.e., away from a high degree of academic, intellectual and scientific activity. The psyche is locating me in the feminine, the earth (the half submerged house) and the pig, in the Berkeley flatland near the water. My soul image, the anima as Jung called it, is presented in this dream as an older woman with her black (Isis' color) sow in a particular context. My Zurich training was much more about the realm of the archetypal feminine—the intuitive, the irrational, the body and the earth, the moist darkness of the unconscious—than about the arid realm of my Ph.D. work in entomology at Berkeley.

This dream was an evolution of an earlier dream I had while working on a Masters degree in Humanistic Psychology at Sonoma State before going to Zurich:

> I'm sitting at a school desk with two or three high school age males. We're waiting to talk to a guru, a sort of wise old man, a sagely type. It's getting late in the day. I'm concerned that I'll not get in to see him. An instructor I know is tending the door for the "wise man." It's a card-board door on a rather small building, almost a hut. Now its 5 p.m. I don't get to visit the guru and I'm disappointed. The doorkeeper says: "He would probably just tell you that you already know everything there is to know." I go off to the left. There is a dark colored wooden shack with the front exposed to the elements. The earthen floor is sunk below street level. Two women my age are within. One looks poorly nourished with tattered clothes darkened with dirt and grime. It seems this is where the real learning is for me.

The doorkeeper was an instructor at Sonoma State. He was an intellectual, strong thinking type, and an expert on Rollo May, one of the "big three" of humanistic psychology's "founding fathers." His remark in the dream stuck me as one of those clever things a self-described "enlightened" intellectual would say. The alternative to this type of male spiritual knowledge was offered by the two women in their type of abode. The common theme with the woman-pig dream is clear. Being a strong thinking type myself, the message from the psyche was also quite clear.

Just a few more pig impressions before looking at their favorite food—corn. They're quite talkative animals, especially if compared to cows. They're often grunting, "uffing" is a better word, letting each other and you know they are aware of your presence and what they think about it. They love to be stroked behind the ears but hate young boys riding them. At least it's not like falling off a cow's back. What's most unnerving is a pig's squeal. When they crank it up full volume it strikes primal fear in the psyche. You get a modulated version of that sound if you try to get a pig to go where it really doesn't want to go. Being so big and close to the ground, they're not easy to move and they complain loudly. Anybody who has raised pigs knows all too well what it means to be "pig headed." The pig is accused of being "intractable" and "difficult" in hexagram 61 in *The I Ching*, called Inner Truth:

> Pigs and fishes are the least intelligent of all animals and therefore the most difficult to influence. The force of inner truth must grow great indeed before its influence can extend to such creatures. (Wilhelm 1967, p. 235, 236)

The pig as an emblem of ignorance or unconsciousness appears at the center of the Tibetan World Wheel with two buddies—the cock as the sign of concupiscence and the snake as a symbol of hatred or envy. The wonders of scientific measures of animal IQ's tell us that pigs are actually smarter than cats or dogs. They're really quite clean animals (so I've been told) and only wallow in mud because it's cool, not because it's cool to be uncouth.

The total impression of pigs would be incomplete without commenting on their smells. There is nothing worse than the smell of a pig farm. When driving through Iowa on the Interstate, you can tell when you're within a mile of such an establishment. I'll take the smell of cow manure any day.

A tour through the Midwest will balance the psyche between the world of animals and the world of plants. The plant environment is dominated by corn, wheat, alfalfa and soybeans. To imagine what it's like to be a plant, you must cultivate a vegetative imagination. We're talking about being rooted, grounded, firmly anchored in the earth, otherwise you're washed out, washed up, blown away. Only as a rooted plant can you have that intimate connection to the earth that provides life with it prime mineral supply. Yours is a life of rootedness and absorption—absorbing minerals and water through the roots and CO2, life's waste gas, through your leaves. The greatest miracle of all, and upon which *all* life depends, is absorbing sunlight. Only you can capture the emanations from the sun and use that energy to drive a chain of chemical reactions that combine water and CO2 to create sugar—the molecular foodstuff and energy source of life. Plants also manufacture most of the amino acids needed by all other living organisms to build the enzymes and other proteins that make each body unique.

So when you pass that corn field, wheat field, soybean or alfalfa patch, don't think "motionless" and simply "existing." Think "absorption"—absorption of dense mineral matter and water from a cool, moist earth; absorption of air and sunlight through delicate leafy appendages. Think manufacturing of life's building blocks and energy sources. Think oxygen, plant's by-product gas of life, all produced by staying rooted in one place and absorbing the environment.

Contrast vast acreages of corn, wheat or soybean monocultures with alfalfa, which implies a dairy farm and therefore a variety of crops to feed the discriminating cow palate. Think drilled and lined up in rows versus a solid field of alfalfa as thick ground cover. Grant Hill's painting, "Stone City Iowa," epitomizes the buttoned down, grid-over-the-land feeling one gets from row crops on a landscape. Also think wind pollinated (wheat and corn) versus bee pollinated (colorful flowers and the beautiful fragrance of alfalfa). Cornfields have their own distinct smell, and driving past cornfields on warm summer nights will yield a windshield full of smashed moths whose larvae love corn as much as pigs do. Think of being lost or disappearing into tall corn plants versus being in a waving mass of sweet-smelling, thigh-high alfalfa plants.

Corn deserves special attention because it rightfully has been called the quintessential American crop. It originated in the Americas, Mexico by best estimates. Within twenty years after Columbus "discovered" America, the Spanish and Portuguese sailors had spread it to Africa, the Middle East and shortly after to China. (Fussell 1992, p. 18) It quickly adapted to the varied environments of its newfound homes and transformed barren lands into fruitful cornfields. It accounted for a population explosion in China that began in the seventeenth century. (p. 163) This "hopeless monster" as geneticists call it is endowed with a wildly variable gene pool that allows it to adapt, through natural and artificial selection, to a wider range of growing conditions, environments and industrial purposes than any other grain. It is unsurpassed in energy conversion, yield, speed of growth, and edibility of most of the plant. (p. 20)

This strange plant is totally dependent on humans for its reproduction: the husk wraps the seeds too tightly to disperse; if a shucked ear is buried, the young shoots die of overcrowding. (Fussell 1992, p. 20) The tremendous increase in corn yields (from 25 bushels per acre in the 1920's to 125 bushels today) comes from cross-pollinating (by hand!) various strains of corn to produce hybrids. (n 2) Farmers can't plant their harvested corn seeds, making them dependent on seed producing companies to start each hybrid generation from scratch.

Corn is one of the "big three" grasses that were domesticated into grains between 9000 to 5000 BCE, with corn originating in the Americas, rice in Asia and wheat in Mesopotamia. This changed the previous two-million-year course of humanoid, then human hunter-gatherer societies into cultivators of crops. Agriculture allowed humans to settle down, multiply, and gather into population centers which permitted the rise of complex cultures, the arts and civilizations. (Fussell 1992, p. 39, 40)

The famous pre-Columbian American civilizations—the Aztecs, Incas and Mayans—were built around corn. Corn became the central object of religious worship and integral to their language and calendar systems. (see Appendix B: Sacred Corn) The Anasazi of the American Southwest and their Pueblo descendents were the northern cultural derivatives of old Mexican corn cultures. The Hopi, of Anasazi descent, influenced the Navaho, one of whom described the sacred relationship with corn like this:

> When a man goes into a cornfield he feels he is in a holy place, that he is walking among Holy People, White Corn Boy, Yellow Corn Girl, Pollen Boy, Corn Bug Girl, Blue Corn Boy, and Variegated Corn Girl...If your fields are in good shape you feel that the Holy People are with you, and you feel buoyed up in spirit when you get back home. (Fussell 1992, p. 95)

Corn played an important role in many North American Indian tribes. The Pawnee, for example, associated corn with the Evening Star, "the mother of all things, who gave corn to the people from her garden in the sky." (Fussell 1992, p. 16)

One reason corn is quintessentially American is because indigenous peoples of the Western Hemisphere originated and developed the largest number of strains. Of the roughly 280 races of corn, 210 are unique to South America, mostly to Peru, and 30 to Mexico. (Fussell 1992, p. 87) But it is the Americans who are noted for the tremendous development in agricultural practices and the scientific and technological manipulations of corn's genetics and raw materials. The quaint Mitchell Corn Palace in South Dakota is the emblem of a tradition that trumpeted the union of science, commerce, agriculture and art that made corn "the ground of the entire industrial empire that sprang from the prairies and plains." (p. 313) (n 4) Americans directly eat only about 1% of the corn grown here (n 3), preferring to use about 85% to fatten cows, hogs and poultry because its powers of conversion of food to meat is double that of wheat. (p. 7) We enjoy drinking the distinctly American oak barrel-aged corn liquors known as Jack Daniels Whiskey and Old Crow Bourbon, and the not so famous un-aged corn liquors of moonshine fame.

We Westerners have used our scientific and industrial ingenuity not to make recipes for eating corn but to develop a staggering array of products from corn kernels. Following the motto of the corn synthesizers, "anything made from a barrel of petroleum can be made from a bushel of corn" (Fussell 1992, p. 265), the corn molecules have been twisted and tortured into an incredible variety of products that touch almost every aspect of our lives. Corn products are hidden in antibiotic components, soaps, medicine tablets, plastic cups, and drying powders in ready-mix cakes. They are used in making bread, the bodying agent to improve the "mouth feel" of chewing gum, and as food preservatives. Other uses are in toothpaste, an emulsifier, superglue, and in hard, light-weight building materials. We drink more than enough corn in the form of sweeteners (sugars derived from molecules of corn starch) used in the mammoth soft drink industry. (p. 269-277)

The complex and broad dimensions of ecopsychology are seen in issues concerning corn. Making ethanol from corn is being touted as a response to the oil barons of the world and a food-or-fuel controversy is brewing. International water wars will be fought out in the grain commodities markets as developing water shortages for crops, especially in India and China, will force grain purchases on the international markets. (see Lester Brown *Plan 3.0*) China holds over a trillion dollars of American debt from the unfunded Iraq and Afghanistan wars and may not take too kindly to being denied grain purchases when their people begin to starve.

The contrast between the indigenous and the modern attitudes towards corn best illustrates the split in the psyche's relationship to nature. Natives are more concerned with the symbolic meaning of corn than with productivity, maintaining pure strains of color and type because of sacred meanings for each color. It is they who developed the multitude of inbred strains. (Fussell 1992, p. 65) The "industrialization" of corn breeding (hybrids of these inbred strains) and mass marketing from the 1920's to 1950's laid the foundation for modern American agribusiness. (p. 67) Its ugly by-products include soil erosion and depletion plus fertilizer run off and contamination of waterways. The NAFTA trade agreements resulted in a flood of cheap, subsidized American grain into Mexico, driving small farmers off the land. They become the slum dwellers in huge Mexican cities and illegal immigrants in America desperate to survive.

We arrive at our last essential image of the Midwest, a natural derivative of a gestalt of environmental factors. That is, of course, the farmer, the one who cultivates and harvests the rich Midwestern land. Thomas Jefferson envisioned America as a nation of yeoman farmers. Hilgard Hall, the old agriculture building where I did my research in Berkeley, had this challenge chiseled into a wall: "To rescue for humanity the innate values of rural life." The Midwest is the keeper of our nation's self-image of having rural values, now highjacked by the term "family values."

Exactly what are these values? Consider the farmer: independent because he makes a living off the land, has no boss and is beholden to no one; is therefore democratic, not elitist. He's pragmatic and a jack-of-all-trades. He's wholesome because he lives and works in nature; not decadent, lazy, without drive as people in cities are imagined to be. He is humble because he depends on nature and is subject to its vagaries; is open, moral and more innocent because away from big city life. I'm not saying farmers are basically like this, but there are kernels of truth in each of these statements. (n 5) I used the word "he" for the farmer, but having seen how long and hard my mother worked on the farm, I've never questioned the strength, stamina and intelligence of a woman. (n 6)

The down side of the farmer image is hayseed, country bumpkin, dirtball, and closed-minded. A farmer may also appear to be materialistic and unimaginative. Sinclair Lewis' *Main Street* was the first to deflate the idealization of the innate values of rural life and small town America. (Shortridge 1989, p. 45)

The obvious industrial development of the Midwest with cities like Chicago, Detroit and Cleveland challenges our basic image of the Midwest. Agribusiness and huge corporate farms further erode the ideal of the yeoman farmer, but studies have shown that Americans cling to the agricultural image. (Shortridge 1989, chapters 4, 5) A *Time* magazine cover story on Minnesota (1973) offered a Jungian anima description of the Midwest: "California is the flashy blonde you like to take out once or twice. Minnesota is the girl you want to marry!" (n 7)

So, our archetypal images of the Midwest are that of being central, heartland, pastoral, cultivated land and farming. It's of farmers; men and women and children working together with nature to produce the food of life in the breadbasket and corn belt and dairyland of America. We think cow—grazing, ruminating, ungainly and docile; a mobile milk factory, a nourishing Great Mother image. Or we think pig—earthy, fertile, meaty; primal side of the Great Mother. We see corn, wheat, soybeans and alfalfa—vegetative life—rooted, absorbing; producing oxygen for the breath of life and foodstuffs for humans and their domesticated animals. We think gentle, rather flat topography given some relief by glacial effects, reminders of eons past and the uncertainties of climate change. We think of the water of life; rivers, lakes, ponds, marshes and bogs. The fundamental image of the Midwest is that of the Great Mother in her fertile and nourishing aspect, of Demeter's largest domain of good agricultural land on the face of the earth. This domain is given a spiritual dimension by the extremes of the continental climate that ravages the Midwest as we will see in the next chapter.

CHAPTER 2

Seasons of the Soul

"Everybody talks about the weather but nobody ever does anything about it." Mark Twain's gardener was right; the weather is an autonomous force we can't control. We can only live with it and adapt to it, developing an I-Thou relationship in the process of coming to terms with it. If we can embrace the weather and open ourselves to it, it becomes a meaningful experience as we gain consciousness of the effects it has on our psyches. If we can't do that, an extreme continental climate like that found in the Midwest will drive us into fighting the weather, forever complaining about it.

The autonomy of the weather and our own reactions to it go to the heart of the dilemma facing Western culture in relationship to the environment. The essential problem is: *the primary myth that our Western psyches live by is the myth of the hero.* The hero lives by conquering and subduing, letting nothing stand in his path—and it is usually a he. It is us *against* them, and *me* against it. This attitude feeds the subject-object split that has created a dead universe of things. *My* subjective psyche, *my* reflections, *my* inner-most thoughts and feelings are all that are alive and where the spirit dwells. The universe is only alive to the degree I project my unconscious contents onto it to enliven it, including projections onto other people. (Hillman 1992, p. 26-37, 98, 99)

Jungian analyst James Hillman suggests the intriguing antidote of re-establishing the neo-Platonist idea of Aphrodite as the Soul of the World. (Hillman 1992, p. 89-130) (see Appendix K of *The Dairy Farmer's Guide*, volume 3) With the Greek goddess of sensuality, beauty, love, the smile and allurement as the Soul of the World, the now dead objects become sensually enticing subjective beings with a sacred dimension. The world becomes a soul-full place, soul being anything that can allure, captivate and move the psyche. Presently the soul is locked up in our individual heroic egos. Breaking this literal ego vessel allows the soul to re-enter the world around us and have it animate us once again, from the outside in, having it come alive in a manner akin to how indigenous peoples see the world. (p. 101-106) We form a deeper connection with nature—a goal in ecopsychology.

Hillman reminds us that the dead universe of matter is coming back to life for us via its pathology, analogous to the reality and autonomy of the human unconscious that Freud discovered through human pathologies. (Hillman 1992, p. 97, 98) The decaying environment impresses itself upon our psyches instead of the other way around. The "dead *things*" are suffering: acid rain, droughts, raging fires, Katrina; the environment is getting our attention. It is showing us it can and will affect us. It's not just our subjective sense, our projections, that the environment is suffering. The very real pathology in the environment generates pathos and empathy, a connection we feel with it. Either Aphrodite lures us into an enticing relationship with nature and a naturally ensuing aesthetics and ethics or we will be dragged by a karmic fate that comes from violating science's ecological laws.

Our attitude towards the environment is analogous to our attitude towards our dreams. They are not *our* dreams; our conscious egos do not make dreams. No, *we* are dreamt. Our dream ego is only one perspective, one character in the dream landscape. That is what Jung meant by the objective psyche. We are in the dream, the dream is not in us. The characters in the dream are autonomous, like little people within. We think of ourselves as a unified, monolithic ego, but the ego is only one perspective, one metaphor among many, within an individual psyche. We must wake up to the reality of our dreams. (see Appendix K in volume 3 of *The Dairy Farmer's Guide*)

I. Weather Effects on Body and Soul

Our dream ego in relation to other elements in the dreams is similar to our I-Thou relationship to weather and the seasons. Over 30% of people are weather sensitive, many clinically so, with all other people being unconsciously affected by atmospheric conditions. (n 1) We are directly affected by the Other in our moods, perceptions and physiology reaching down to the glandular level. There appears to be a kind of diffuse, comprehensive, whole body response to certain environmental influences, especially electrical and magnetic influences. (Redgrove 1987, p. 92) Ionic currents in the air directly stimulate the nerve endings in the skin, thus affecting touch. Ions are absorbed by the lungs—electricity in the very air we breathe—which influences respiration and blood-pressure. (p. 93-95) (n2)

Weather effects are most noticeable in comparing the mental and physical states before and after storms, lending credence to the Romantic notion that the moods of nature coincide with human moods. (Redgrove 1987, p. 82) Biometerology, the study of the effects of weather on living organisms, lists a host of illnesses that increase during the warm and humid conditions that precede a weather change. These include hemorrhages, asthma attacks, migraines, myocardial infarcts, colic, angina pectoris, osteoarthritic complaints and thrombo-phlebitis. Thunderstorms are particularly potent in altering the senses of smell, taste, hearing and touch. (p. 81) People feel stimulated, inspired and uplifted near the end of a storm and sufferers of many diseases feel relief in their aches and pains when the storm is over. (n 3) The passage of air masses affects the pulse and breathing rates, blood pressure, blood composition and other bodily functions. (p. 79)

Many Romantic artists, such as Coleridge and Goethe, were extremely conscious of the affect of weather on their mental and creative states. (n 4) Mistreatment results when the medical community does not recognize weather-related mental and physical problems. (n 5) People naturally will get depressed if this knowledge of their deep connection with nature is withheld from them. (n 6)

Paying close attention to weather and the seasons and appreciating its effects on the psyche is a soul making process. Recognition of the weather is reflected in how we dress for it. A nature guide told me, "There is no such thing as bad weather, only inappropriate dress for the weather." Not to dress properly for the weather is not to be in proper relationship to it; it's not respecting its presence and "givenness." If you catch yourself fighting the weather and climate and complaining bitterly, examine yourself for the possibility that you may have a heroic ego stance towards it—and to life.

I do several things to develop an appreciation of weather and climate. First is to study the weather and seasonal analogies in the *I Ching*. The oldest level of the text was written in 1150 BCE from oral traditions that go back untold generations before that, back to shamanism in northern China. One can put a question of personal concern to the book, like an oracle, and often get an answer presented as an analogy or a metaphor about the weather or a season.

The second thing that helps me "see" the weather and sense its psychic meaning is to use a little trick. When I leave the front door of my house, I often pause and ask myself, "If this weather was in a dream, how would I interpret it?" The weather immediately changes from a purely physical phenomenon (its Maya aspect) to something with depth and meaning.

The third thing I do is take micro-vacations. If I'm walking and something beautiful or unusual catches my attention, I tell myself that people pay good money to go on vacations to experience something this beautiful. This justifies stopping in my tracks and soaking in the beauty of the moment. This may include a quality of light, a play of light and shadow, a beautiful scent, an interesting cloud formation, or the quiet beauty of a snowfall.

The fourth trick I occasionally use is to close my eyes as I walk in the evening. Then I walk more slowly and other senses besides sight come to the fore. Sounds and smells, and the ineffable sensualness of the air: these are particularly powerful on warm summer evenings. One of the first things I missed when I moved to the Bay Area was a warm summer evening. Mark Twain said, "The coldest winter I ever spent was summer in San Francisco."

II. Clouds

I will use many contrasts between Californian and Midwestern weather to differentiate and thereby become conscious of what is so powerful about Midwest atmospherics. One major difference I noticed was the incredible variety of weather in the Midwest in contrast to California. We have almost the entire range of weather types here: it's a weather lover's heaven. I found weather in the Bay Area to be rather boring. In my dream about a typical Midwestern landscape (chapter 1), even the atmosphere and the clouds in the dream were numinous (had a sacred dimension). One thing I missed in California was the Midwest's clouds.

So lets begin in the clouds, with the imagination, with a dream I had about clouds:

> In a silhouette is the side of a mountain with a ledge upon which are standing four people. Now I am one of the four people looking at a beautiful painting with the painter my

> age directly across from me, a beautiful unknown woman to his right (my left) and an unknown man to my left on my side of the painting. The painting consists of concentric oval shapes with colors not unlike the mother of pearl inside a clam shell. As I am marveling at the incredible beauty of the painting, I realize something even more profound: the artist did the painting by arranging pieces of cloud. As I am becoming aware of this deeper level of creativity, the dream scene switches to me walking around in a city and wondering what city it is. A voice says, "Madison."

The four dream figures represent the painter as a personal archetypal image of the Self, the woman as the anima, the unknown man as the shadow, and the dream ego. The dream illustrates Hillman's imagination of Aphrodite as the Soul of the World. The image, the painting, is beautiful: the soul needs beauty; beauty within, beauty without. The image allures, draws one in, enticing one to attend more closely to it as it stirs the psyche. One is transported to a deeper level, what the Iranian scholar and mystic Henri Corbin called the "imaginal level," the in-between world of the gap in Hermes' wand (volume 3). It amplifies and extends the senses, giving texture and depth to images. It is the psychic sense beneath the material level of sensing.

We must see through the illusion that the material realm is all there is. The deeper reality is constructed out of clouds/metaphors arranged as if by an artistic Creator—not a scientist or a technician. The appropriate response is a gasp, a breathing in of wonder, an aesthetic response, not an analytic, developmental description and intellectual concepts. What is called for is an imaginative sensing with the heart as Hillman would say. (see Appendix K in volume 3)

My fascination with clouds goes back to childhood. As an adult, I have to remind myself to look up, and not be ashamed to wonder at the austere beauty and fantasy images in the clouds. As mentioned in the preceding chapter, it doesn't take much earthly intrusion into the sky to keep our attention focused on the earth. The prairies of Illinois offer a flatter, more unimpeding topography for cloud lovers than do the hilly terrains of Wisconsin.

Cloud watching stimulates fantasy and the poetic imagination. On a vision quest one comes to honor them as messengers by the images they bring. Teachers do writing exercises by having their students go out, lie on their backs, and watch the clouds. After their fantasy life has been sufficiently stimulated, they write imaginative prose and poetry.

Earth and sky form a living unit in the Midwest. The greenery here depends on clouds bringing the water of life from the skies: by and large we don't irrigate. The earth gives back water to the sky by evaporation from its surfaces and transpiration from its plants. By attending to the sky throughout a hot summer morning, one can literally see the earth-sky interaction. The day may start out clear, but by noon the sky may be full of puffy white clouds. They haven't just blown in; they formed from the water that evaporated from the earth and plants below: heat, earth, plants, water and cold (producing condensation) have interacted to form patterns in the sky.

The most exciting clouds are storm clouds. I missed thunderstorms when I lived in California—really missed them. The best place to experience a storm is on the prairies. Monster storms can be seen approaching from miles away, a black being creeping over the horizon with a frizzled halo of lightning flashing like a pinball machine. The storm approaches and gradually engulfs one in its primal fury, reminding me of the base guitar lick on "The Narrow Way: part two" in the Pink Floyd's *Ummaguma* album; a shamanic expression of controlled intensity. The *I Ching* describes the impact of thunder and lightening on the psyche by making an analogy to the power of music. In hexagram 16, Enthusiasm, in the section called the Image it says:

> When, at the beginning of summer, thunder—electrical energy—comes rushing forth from the earth again, and the first thunderstorm refreshes nature, a prolonged state of tension is resolved. Joy and relief make themselves felt. So too, music has power to ease tension within the heart and to loosen the grip of obscure emotions. The enthusiasm of the heart expresses itself involuntarily in a burst of song, in dance and rhythmic movement of the body. From immemorial times the inspiring effect of the invisible sound that moves all hearts, and draws them together, has mystified mankind. (Wilhelm 1967, p. 68)

The mythic dimensions of a thunderstorm in the imagination of Native Americans and the sense of the earth-sky connection is presented in this description of The Thunder-beings:

> The Thunder-beings make up the love call of the Sky Nation. The Fire Sticks, or lightning bolts, are a rare gift from the Sky Father to the Earth Mother. The Thunderers who accompa-

> ny a storm carry the mating call that announces the Divine Union of Earth and Sky. The Thunder-beings are the host of lovers who give energy to the Earth Mother. The Thunder Chief proclaims the beauty of the love between Father Sky and Mother Earth. The Fire Sticks create a bridge between the two lovers and are a physical expression of their love for one another. The Cloud People gather where the dance of union is to be held and house the Thunder Chief and Fire Sticks within their bodies awaiting the joyous time.
>
> Through this intricate mating dance, our Earth Mother is re-energized so that life may continue through the nurturing Rains who feed her body. (Sams 1990, p. 267)

The Lakota Sioux associate the West with Thunder-Beings. They traverse the planet, destroying negative forces and old forms with lightning bolts, allowing new forms to emerge and grow with the rains that accompany a storm.

Thunderstorms can bring torrential downpours. Cloudbursts in dreams, such as on a house without a roof, can be a metaphor for being overwhelmed by the unconscious, by powerful emotions washing over us.

The Midwest experiences the ultimate storm—the tornado, the most destructive wind in the world. The polar opposites of yin and yang, masses of Arctic dry, cool air or dry mountain air collide with warm, moist Gulf air to spawn twisters. I get most fascinated by the weather when the sky develops an eerie greenish-yellow color and low ragged skud clouds race along under an avalanche of cloud forms. Then I realize I'd better take cover because a tornado may be brewing!

Tornadoes strike fear in the psyche. Take *this* as a metaphor for a psychological state: deadly, local, unpredictable, disintegration by violent twisting or exploding from within. Tornado dreams are fearful implications of psychic doom. Here's an example of the psychic punch of a tornado dream:

> I'm standing in the open as a tornado funnel cloud approaches me. I know I am going to be hit but I'm helpless and can't move.

The dreamer had this dream shortly before his first psychotic episode.

The extremes of the Midwest continental climate depict in the skies the entire repertoire of psychic states. Think of all the weather terms used to describe psychological states: a "sunny" disposition, a "bright" and cheery person, a "stormy" relationship, an "airy" personality, a "foggy" state of mind, a "cold"-hearted person, a "hazy" view of things.

Another aspect of weather and climate making it a natural for depicting psychological states is its Protean nature: it is always changing, ever transforming. The sky is a big Hermetic vessel, mirroring the psyche: "as above, so below" the alchemists proclaimed.

III. Seasons of the Soul

And then we have the seasons. As a child living in southern California I didn't understand why my mother missed the change of seasons of her Wisconsin youth. After moving to Wisconsin at age 6, I wondered why she missed the hot, muggy Wisconsin summer days or the long, *derriere*-freezing winters.

I've since come to appreciate the richness and beauty of each season. Each has a teaching, something basic to tell us about life. Francois Cheng, a professor of Chinese art, said this about landscape painting: "In paying so much attention to the nuances of a landscape going through the changing seasons, the painter is in reality expressing the state of his own soul." (Chen 1994, p. 80) In Ecclesiastes we read:

> For everything there is a season,
> and a time for every matter under heaven:
> a time to be born, and a time to die;
> a time to plant, and a time to pluck up what is planted;
> a timed to kill, and a time to heal;
> a time to break down, and a time to build up;
> a time to weep, and a time to laugh;
> a time to mourn, and a time to dance;
> a time to throw away stones, and a time to gather stones together;
> a time to embrace, and a time to refrain from embracing;
> a time to seek, and a time to lose;
> a time to keep, and a time to throw away;
> a time to tear, and a time to sew;
> a time to keep silence, and a time to speak;
> a time to love, and a time to hate;
> a time for war, and a time for peace. (Eccl. 3: 1-8 RSV)

Joseph Campbell said the most recognizable, universal forms of myth and analogy are of the seasons to the life cycle. These are archetypically described in the *I Ching* as the four aspects of any cycle of change: Spring, Growth, Harvest and Trial—more meaningful terms than Richard Wilhelm's translation as Supreme, Success, Furthering and Perseverance. (Ritsema et al 1995, p. 66, 67) (n 7) The spiritual dimensions of the seasons encapsulated by these four terms are aspects of the cyclical Tao of heaven at all levels: in the macrocosm and microcosm, in nature and in human life, in inter-human relations and in inner development as well as in all spans of time, such as a day, a year and a life span. The concepts are analogous to the attributes of the four directions on the Native American medicine wheel with Spring being the East, Growth the South, Harvest the West, and Trial the North. (fig. 1)

When we consult the *I Ching* about a crucial life issue, we want to know where we are in the developmental cycle of that issue, what season we are in, be it about a job, our family, or personal or spiritual development. The many seasonal and weather analogies given by the *I Ching* make it easier for us Midwesterners to have an experiential sense of the meaning of the *I Ching's* answers.

Spring for the Chinese includes the primal, originating power itself, and the place in space and time where it becomes phenomenal, i.e., the source, beginning, the head, etc. It is the first sign of day, the first season. (Ritsema et al 1995, p. 66) On the medicine wheel of many Native American tribes it would be comparable to the East where the sun rises, hence the color yellow. For many tribes its animal is the eagle. (Storm 1972, p. 6) Eagle soars closest to grandfather sun. It symbolizes freedom, lofty ideals, the light of understanding and the courageous bursting forth past old fears and limitations into new territory. The spark of life and light that is Spring is associated with visionaries, illumination, clarity, imagination and the genesis of ideas. (Joan Halifax interview on "New Dimensions" radio; Storm 1972, p. 26) It is Aries in astrology, hexagram 1 in the *I Ching,* and the thinking function in Jung's typology system. (fig. 1)

Spring in the Midwest begins with the schizophrenic month of March, where warm spring-like days may be followed by blustery, cold, even snowy days. This lion-and-lamb combination presents a great drama to the psyche. It goes through periods of elation and depression as spring comes and goes. Its exaggerated ups and downs are like Cinderella going to the ball, then falling back into the pits. Finally spring is established when tree leaf buds burst open to herald the change; the prince has shoed Cinderella.

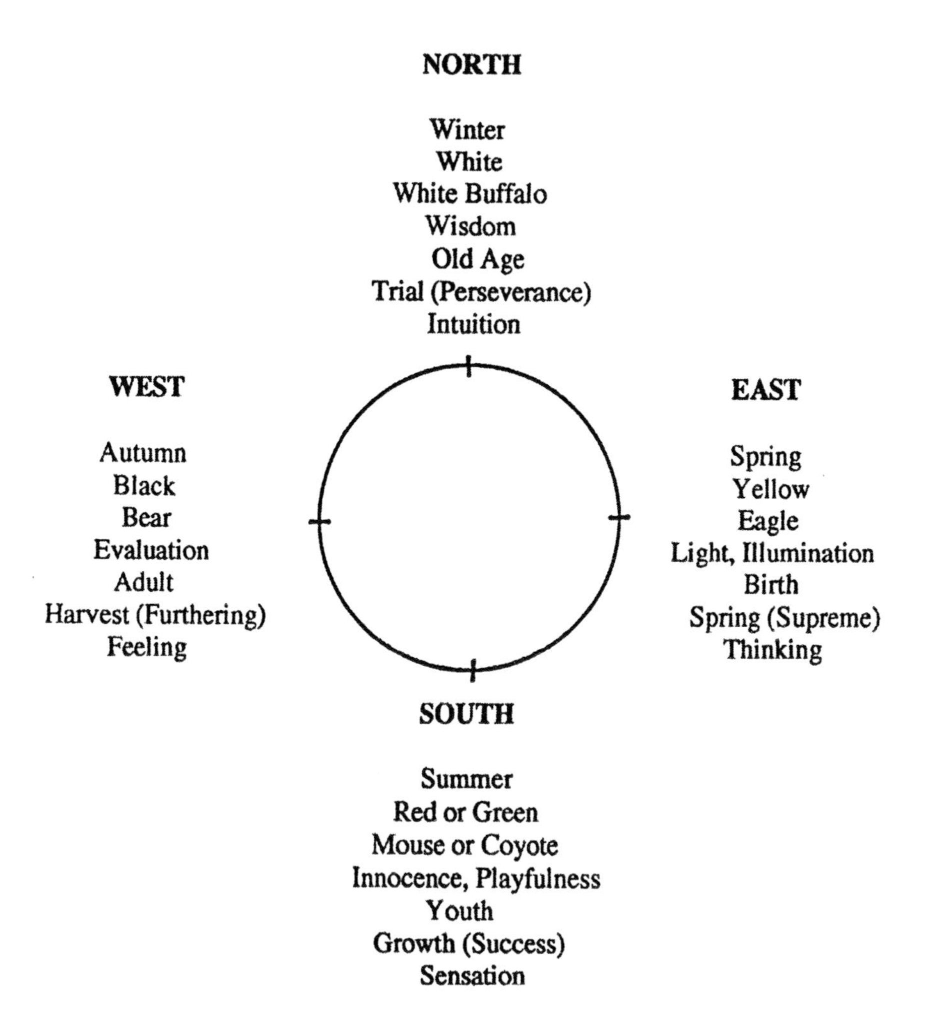

Figure 1. The medicine wheel.

Spring is a bursting forth, a beginning, the moment of coming into existence. (Ritsema et al 1995, p. 66) Just as the sun follows darkness, the first spring flowers burst forth out of winter's dead earth. Bud burst for the trees is my favorite period of the year: trees are painted subtle pastel shades of green, yellow-green, soft reds and pinks as leaves and flowers come forth. It lasts only a few days unlike the weeks of blazing fall colors. One can see the tree forms through the canopy of tiny leaves and admire the glacial topography of the hills beneath.

We hear the sounds of the heralds of spring as geese, robins and red-wing blackbirds return. The succession of spring flowers and the warming days lure us out to embrace nature and experience the child's world of exuberance and liveliness. Life returns with the greening of the earth and trees, and the sensual fragrance of flowers. Winter coats come off, spring romances flourish. For life to flourish, old forms in the earth must be broken up with tilling before the seeds of a new cycle can be planted.

Growth, the summer season, meant to the Chinese the all-pervading extension of the generating principle and for carrying everything through to full-grown forms. The spark, the vision, the idea in **Spring** takes form. (Ritsema et al 1995, p. 66, 67) It is related to South on the medicine wheel; often the colors are red or green. (Storm 1972, p. 6) Red is the color of life's blood, of passion and of faith. The South is the place on the medicine wheel where physical life begins. It is associated with the child. The child celebrates being itself and wonders at life's beauty. There is innocence, trust and faith in life. (p. 6) Playfulness is expressed through the body. The coyote is the animal frequently associated with the South. He tricks and humors us when we become self-deceptive, too serious, too pompous, too self-important. Coyote returns us to a position of humility and innocence. (Joan Halifax interview on "New Dimensions" radio) Mouse is also associated with the South. She pays careful attention to the sensations and details of a close, physical environment (the sensation function) and "our nature of heart." (p. 6)

What was planted in **Spring** must be tended and nourished. **Growth** is the energy of the Great Mother in her nourishing form in summer. Life flows into forms to create the green mantel of summer. Summer: a time of brightness and heat, of activity and physicality, of food and backyard barbecues with friends and family. It's the lazy, hazy, crazy days of summer and summer vacations.

A paradox is felt by the psyche: the longest day of the year, June 21st, the first day of summer, tolls the beginning of the decline of the light. The sense of the seed of darkness planted at the height of light is captured as the dark eye in the white yang of the Chinese yin-yang symbol. (fig. 2)

Summer's heat and light plus water from the heavens provide the elements to nudge life into its mature forms. The seeds of the next generation silently form as crops ripen. The sequence of flowers ends with the goldenrods and asters of August and September. Browns and yellows creep in. The fruits of summer are ready for the harvest, the third stage of a cycle.

Figure 2. Yin-Yang Symbol *t'ai chi tu.*

Harvest in the *I Ching*, the autumnal season, is the moment when the phase of **Growth** is acutely cut off, like a scythe reaping grain to gather in the profits. (Ritsema et al 1995, p. 67) This is West on the medicine wheel when its animal, the bear, withdraws to hibernate in a protective cave, the "Looks-Within Place." (Storm 1972, p. 6) It is connected with archetypal feminine energies, with containment and introspection, insight and emotions. It is the midlife position. The West is the digestive place of feeling evaluations of experiences. It is receptive to answers and directions arising from going into the dark silence within and the strength and courage of the bear to act on directions received. (Joan Halifax interview on "New Dimensions" radio)

The **Harvest** time of fall is the favorite season for many of us in the Midwest. The intense heat of summer is gone, skies are often crystal clear, the weather is crisp and invigorating—it's football season. The bugs get killed off—mosquito season is over! The trees go out in a blaze of glory. Smells are multitudinous and divine—apples ripe or rotting, molding leaves in the woods.

People begin to move indoors. Summer's green mantle is gone. Dead brown leaves abandon trees, leaving woody skeletons behind. Even the geese leave us as we experience the last of the great animal migrations. A certain melancholia creeps in as the days get shorter and the storms get colder and nastier. The psyche feels the throes of death.

Winter drives us further inwards; into homes, deeper into our psyche, deeper introspection. With winter clothing we protect our life force and conserve life energy. A blanket of snow reduces the landscape to black and white. Days shrink until the winter solstice, the shortest day of the year, the first day of winter. The second paradox of the seasons is experienced: when the night is longest, the seed of new light is sewn as the sun will begin its journey north. A spot of light crystallizes in the blackness of yin. (fig. 3)

Figure 3. Yin-yang symbol *t'ai chi tu.*

The sun's promise comes while the depth of winter's oppression lies ahead. The cold, the blizzards, the winter that goes on and on, tries the soul. **Trial**, the winter season, meant to the Chinese the testing by ordeal to produce the pure, undefiled, resistant kernel of meaning, an intuitive sense of the meaning of it all. To inquire by divination is one of the meanings of **Trial**. (Ritsema et al 1995, p. 67)

This is North on the medicine wheel, color white, whose animal is often the white buffalo. (Storm 1972, p. 6) The buffalo provided the Native Americans with food, clothing and shelter, and a white buffalo refers to spiritual providence to be honored with gratitude. North is the place of the wisdom of old age. (Joan Halifax interview on "New Dimensions"; Storm 1972, p. 6) The physical body fails and the spirit grows stronger. The desires and physicality of **Spring** and **Growth** have long since gone. We can look back with the objectivity of time and distance to distill the meaning of the cycle we have passed through. The meaning of it all slowly crystallizes, revealing the inner beauty of winter—the inner light of the crystalline world.

Life has died off leaving elemental forms behind. One can see the topography of the landscape most clearly in winter. A German neighbor helped me appreciate trees in winter; he prefers the winter trees because he can admire their architecture. The invisible movements of Hermes almost show themselves in the paths of snowflakes. If one watches a snowstorm from an upper story window one can see the snow swirl around buildings and sweep through passageways. This is as close as the wind comes to revealing to the human eye its changing directions and patterns.

Winter brings the Persephone-in-Hades phase of the Demeter-Persephone myth. Demeter was the Greek goddess associated with the life of the seasons, the growth of grain, of agricultural vegetation; she taught humans the practice of agriculture. She is the Earth Mother who suffers the abduction and rape of her daughter Persephone by Hades, god of the Underworld. Out of anger she refuses to let vegetation grow, hence the winter months. Persephone becomes Queen of the Underworld before returning to be with her mother in spring, who then allows crops to grow. Jungian analyst Pat Berry describes Persephone in the underworld as a de-literalization of reality, as being able to see the essence and archetypal pattern beneath the literal surface of things, as seeing the depth and the interiority of life and objects. The concrete natural world is seen so deeply that it is experienced as a germination in the invisible realm of Hades, an expression of soul. (Berry 1975) Hades cuts one off from materialism and growth, from sensation and the physical body of Demeter. (n 8) The teachings of winter are to realize the wealth of Pluto, being able to see the underside of life, its darker shades, the shadows that add richness and depth to any photograph. It's seeing death in life—developing objectivity by distancing from the emotional ravages of life. Women enter this phase as a life cycle change at menopause. Middle-age men lose physical and sexual prowess, hopefully to acquire the wisdom brought about by aging. It's the necessary death metaphorically depicted as the death of Christ that precedes a transformation, a new life. There can be no rebirth without a death and Fritz Perls reminded us, *"It isn't easy to die and be reborn."*

Spring eventually comes, even to Wisconsin! The days gradually lengthen as light returns. Warmth melts the rigid, crystalline form of water into its liquid, flowing state. Persephone begins to emerge from Hades, led by Hermes. Snowdrops blossom to complete the cycle for us. The seasons turn.

IV. The Symbolic Dimensions of the American Holidays

Another way to recognize the psychological and spiritual dimensions of the seasons is by observing the types of rituals, festivals, activities and symbols associated with each season. People across time have had similar responses to recurring events in their environment and in their lives, a fact that led Jung to develop the concept of the collective unconscious. Every culture displays a unique expression of the archetypal themes. Full participation in the particular rituals and activities of any one culture deepens the experience of the universal dimension of life. We can find a deeper meaning in the holidays by understanding their origins and psychological bases, and by learning about their history and evolution. We can establish a richer symbolic connection to the seasons that can help us reconnect with the land and provide meaning to the seasonal changes.

American holidays provide a symbolic syntax for the year and guide our psyches through seasonal changes. They help us transition from one season to the next. Many of our holidays are seasonal rites with seasonal symbolism being an important part of their appeal, particularly the association of Christmas with winter and Easter with spring. (Santino 1994, p. 17) To a lesser degree Independence Day is associated with summer and Labor Day with summer's end. (p. 10) Rituals are often associated with the holidays because they serve the important function of facilitating transitions. (n 9) Symbols are the minimal building blocks of ritual (Victor Turner 1967 referenced in Santino 1994, p. 14), therefore a symbolic exploration of the seasons provides us with the building blocks of the seasonal rituals.

Many American holidays are related, directly and indirectly, to the solstice and equinox celebrations in ancient European cultures. (Santino 1994, p. 18) Solstice is derived from the Latin *sol* for sun. Solstice means "sun-standstill": twice a year, at the Winter Solstice on December 21st, the first day of winter, and the Summer Solstice on June 21st, the first day of summer, for three days the sun appears to rise at the same point on the horizon (stands still) at the respective Solstice point.

Between the winter and summer solstices, the sun in the northern hemisphere rises and sets every day further north on the horizon. Between the summer and winter solstices every day it rises and sets further south on the horizon. At a point half way between the longest day of the year on the summer solstice, and the shortest day of the year on the winter solstice, comes the equinoxes, when day and night are of equal length. The spring, or Vernal ("green") Equinox, on March 21st marks the beginning of spring and the Autumnal Equinox on September 21st is the first day of autumn.

The reason for the seasons is the tilt of the earth in relation to the sun. At the summer solstice, the Northern Hemisphere is tilted towards the sun. At the winter solstice, the Southern Hemisphere is tilted towards the sun and the Northern Hemisphere gets less light and shorter days.

I will draw on Jack Santino's excellent book on the American holidays, *All Around the Year,* to present their history and meaning. I add a Jungian commentary to illustrate how the seasonal changes reflected by the holidays can be associated with psychological states, connecting inner and outer reality and enhancing our connection to nature.

Most American holidays have their beginnings in ancient religious and agrarian festivals which preceded Christianity. The actual dates of the celebrations do not correspond exactly to the solstices and equinoxes due to historical factors, the major factor being the evolution of the Western calendar. In 46 BCE Julius Caesar reformed the old Roman calendar, which over the years had gotten out of phase with the natural year, losing time so that the name of a winter month may have occurred in spring for example. (Santino 1994, p. 19, 20) In 1582 Pope Gregory XIII modified the Julian calendar to more accurately align it with the solar year. In doing that he skipped 11 days on the old calendar, going from October 4 on the proclamation date to October 15 the following day. Consequently many of the holidays no longer fell on solstice and equinox dates. (p. 21)

Our American winter holidays and the other special days we associate with winter evolved over a long history from various strands of our European stock, and the Church's attempt to put pagan practices into a Christian context and interpretation. A rich amalgamation of these practices then occurred in the United States with some uniquely American contributions added to the mix.

Winter is associated with death. It begins with the killing frost in autumn that symbolically strikes at the life force in our psyche. Life forms disappear as birds fly south and animals retreat into hibernation or are killed off (the insects). There is a steady decline of energy as the days get shorter and colder. The environment becomes hostile to life and survival. Winter holidays and special days are therefore associated with death, and a subsequent bolstering of the life force so it can survive and flower in the spring.

Halloween is the seasonal event most associated with death, when people play with death as they fear it. The roots of our Halloween events go back to the pre-Christian celebration of the first day of the Celtic New Year, the beginning of winter for the Celts. The Celts once inhabited much of Europe before being pushed to the hinterlands by the time of Christ. Present day descendents are the Irish, Welsh, Scots and the Bretons of northwest France. November 1st, called Samhain (pronounced Sahwen), may have been the most important Celtic feast day; it was a focal point of the year: "battles were fought, journeys begun, wars decided." (Santino 1994, p. 148)

The death of the vegetation was associated with the death of the old year—and human death. In many primal cultures the King, associated with the entire fertility of the land and the tribe, was actually killed as part of a seasonal ritual. The Celts believed "the souls of those who had died the previous year assembled [at Samhain] to travel into the land of the dead." (Santino 1994, p. 148) Samhain was filled with the supernatural because the door to the underworld was believed to be open at that time, and spirits and creatures could come over from the other side. Bonfires were built to light the way for the wandering dead to find their way into the underworld as well as to scare away the creatures from the other side. Fruits and vegetables were sacrificed to honor the dead, expiate their sins, and appease the spirits. The night was dangerous. (p. 148, 149)

The unconscious has always been associated with the realm of the dead and the ancestors. The first descent into the unconscious is the most fearful because our shadow side, our dark element, predominates at that time. The ritual descent at Halloween helps a person and a culture deal with all the death in the environment. That constellates or causes to emerge (complexity theory) in the psyche the archetype of death.

The festival of Samhain began at sundown on the evening prior, October 31st, and was associated with the end of the harvest. It marked the end of the crop farming year and cattle were moved indoors. Excess livestock were slaughtered due to a lack of food to carry all of the cattle through the winter, so it was a time of great feasting. (Santino 1994, p. 148, 149) This makes for a strong association of death with the enrichment of life.

One Celtic saga describes a race of supernatural beings who demand tribute from the native people. People delivered goods and harvest fruits to them at Samhain. "[This] parallels the folk custom of setting out food and gifts to appease wandering spirits [and another] folk practice of giving gifts of food and drink to maskers who imitate those spirits." Our Trick-or-Treating at Halloween is a remnant of this practice. (Santino 1994, p. 149) The "trick" part represents social inversion typical of Halloween and other festivals where societal rules are suspended. (p. 151) Celtic tales of fairies and witches were influenced by Nordic myths about "spirits and deities who flew through the air to gather souls and reward heroes." (p. 164)

These beliefs and practices were particularly objectionable to the Church, which saw them as manifestations of the Devil. Christianity transformed what had been powerful, dangerous trickster energies into something imagined to be deliberately evil and malevolent. The Celtic underworld became hell and the Druid priests were labeled devil worshippers. (Santino 1994, p. 153) A half million or more people were put to death in Great Britain and Europe in the sixteenth and seventeenth centuries for witchcraft. Witchcraft was associated with pre-Christian and non-Christian practices and religions, such as festivals, dancing, revelry, shape-shifting and nocturnal flight. (p. 158) Nature, herbal healing, and the old Greek God of nature, Pan, became associated with the Devil. (p. 157)

The Church tried to shift the pagan practices of Samhain into a Christian context by declaring November 1st to be All Saints Day, and November 2nd as All Souls Day to honor everyone who died the previous year. (n 10) This was part of a practice initiated by an edict of Pope Gregory I in 601 AD. He instructed the missionaries to use the native customs and beliefs for conversion to Christianity rather than trying to obliterate the practices. If pagans worship a tree, for example, "consecrate it to Christ and allow them to continue their worship...Catholic holy days and feast days were set at the times of native holy days, festivals and celebrations," writes Santino. (Santino 1994, p. 153)

"Hallow" means "Saint." The night before All Saints Day was called All Hallows Eve, hence the name Halloween. People continued an intense celebration of this evening "as a time of the wandering dead... [and] supernatural beings...now associated with evil. The folk continued to propitiate those spirits and their costumed, masked representatives with gifts of food and drink." (Santino 1994, p. 154)

The potato famine had driven massive numbers of Irish to America in the mid 1800's at a time when the Celtic cultural unconscious still resonated deeply in them. (Santino 1994, p. 153) The basis of the American Halloween was provided by Irish beliefs and customs that merged with Puritan beliefs in witchcraft. (p. 159)

Our Jack O' Lantern is a direct descendent of a legendary folk figure in Great Britain:

> Jack O' Lantern is said to be the spirit of a blacksmith named Jack who was too evil to get into heaven but, because he

> had outwitted the devil, was not allowed into hell. Turned away, he scoops up a glowing coal with the vegetable he happens to be eating, and uses it as a lantern to light his way as he wanders the earth. (Santino 1994, p. 156. 157)

The Irish and British carve large turnips into jack-o-lanterns. We Americans use a native vegetable, our beloved pumpkin. (p. 156)

The custom of bobbing for apples may go back to a harvest festival in Roman times and the Romans had conquered much of Northern Europe by 50 BCE. On Pomona night, a night of mystical revelry in November, they laid out apples and nuts in tribute and thanks to this goddess of gardens and fruits. The apple was closely associated with the goddess because when cut in half it revealed a 5-pointed star, a symbol sacred to the Great Goddess. (Walker 1983, p. 48-50).

Colors are also symbolic of particular seasons. Black, orange and yellow are the colors for Halloween: black for death, and orange and yellows for the leaf colors, fall flowers (marigolds) and pumpkins. (Santino 1994, p. 67) Besides images of death and spirits, Halloween decorations include harvest imagery—corn shocks, gourds and pumpkins. (p. 162)

More than any other national celebration, our costumed Halloween parties allow adults to reveal and play with the dark, shadow sides of their personalities and of life. Halloween is a popular folk holiday which people celebrate despite not being given a day off. (Santino 1994, p. 167)

While the Celts feasted on October 31 in the Samhain harvest celebration, the Jewish harvest festival, the Sukkoth or Feast of the Tabernacle, had been established by Moses and occurs in September or October. The ancient Greeks and Romans had harvest festivals in October with the Romans honoring Ceres, the Roman goddess of wheat in a harvest festival called Cerelia. Our word "cereal" is derived from this name. Festivals were held throughout old Europe associated with the grain harvest. From the Middle Ages to modern times, English countrymen would name a village girl the Corn Maiden, "adorning her with a wreath made from the cuttings of the last sheaf of grain ["said to house the spirit of the Corn Mother"], and parading her through the fields and town." The Corn Maiden was an aspect of the English Harvest Home tradition where feasting "was part of the celebration of bringing in the last crops of the field." This English celebration is the likely historical forerunner to our uniquely American harvest festival—Thanksgiving. (Santino 1994, p. 168, 169)

Thanksgiving has an historical and political twist that gives it an interesting archetypal spin after beginning on a troublesome note. It originated with the Pilgrims who had an October harvest feast day with the Indians in 1621 to celebrate the survival of their first difficult year in America. Survival was due largely to Indian help and generosity. Our foundation story ignores the template established by the Pilgrims of betraying the indigenous peoples and the wisdom they embodied in their sacred connection to the American soil. Two years after the first Thanksgiving the Pilgrims held a somber religious day of fasting, prayer and reflection. Before the end of the seventeenth century, feasting and giving thanks were combined into what we now know as Thanksgiving. (Santino 1994, p. 169, 170)

"Thanksgiving remained the primary holiday celebration of New England for two centuries," notes Santino. President George Washington proclaimed Thursday, November 26, 1789, to be a national Thanksgiving holiday. (Santino 1994, p. 170) Washington called for a commemoration of the Pilgrims and thankfulness for the success of the war of Independence, the establishment of the Constitution, and the well-being of the new country. (p. 170-172) This put the event into the more archetypal domain of founding fathers and founding principles, that is, in the realm of the ancestors. The late November date was too late for it to be a harvest feast in a direct way.

The holiday grew in importance until 1863 when President Lincoln proclaimed the last Thursday of November to be the national Thanksgiving Day, thus ending regional variations. By doing this in the middle of the Civil War, Lincoln hoped to emphasize the common heritage of our country, again an emphasis on ancestral roots. (Santino 1994, p. 172)

The day is associated with giving thanks for the great bounty of the country. From the beginning, America has been imagined as an Eden of boundless fertility and abundance, especially favored by God. (Santino 1994, p. 173) This is symbolized by the excessiveness of the Thanksgiving meal and the super-sized, All-American bird, the turkey. Our link to the land is emphasized with the indigenous American foods of turkey, corn, cranberries, pumpkins and potatoes.

We remember our origin myth about our Pilgrim forefathers at Thanksgiving. The Pilgrims are often depicted as a male and female pair—an American Adam and Eve. (Santino 1994, p. 175, 177) We return home for the holidays, to the home of our parents or grandparents, making Thanksgiving the most traveled holiday. Family values are felt—togetherness, solidarity, community. Elder energy is honored because it is that which survived, sustained, and passed on life to us. Santino emphasizes, "The focus of this holiday is on tradition and continuity." (p. 176, 177)

Whereas Halloween emphasizes death and the fearful shadow met with the first exposure to the unconscious, Thanksgiving taps into the self-guiding and reinforcing energy of the unconscious depths. The archetypal determinants of human nature, with the ancestors as guiding spirits, is seen in contemporary Japanese society. When asked why the murder rate is so incredibly low in Japan, many people responded by saying, "our ancestors are watching us." Christmas on the other hand will emphasize new life and the children of the future generation. (Santino 1994, p. 177)

Gray, yellow and brown are the colors of Thanksgiving: gray the color of the bleak November skies and yellow and brown of a pre-snowy landscape. These colors are mirrored in the dull Pilgrim's garb and the roasted turkey. (Santino 1994, p. 67)

Contrast this with the colors of Christmas—the red of holly berries and Santa's suit together with the green of evergreens. The red of life's blood and the eternal green of vegetative growth symbolize the hoped for rebirth and renewal of life that we need to sustain us through the long dead of winter. (Santino 1994, p. 67) The color white mirrors the snowy landscape that at a deeper level symbolizes the purity of a new birth, as "pure as the driven snow."

The actual date of Christ's birth is unknown. Church fathers picked December 25th over 300 years after Christ's birth, a date popularly observed for centuries. Rome heavily influenced Christian ceremonies because the Holy Land was part of the Roman Empire at the time of Christ's birth and in the early centuries of Christianity. The Roman Saturnalia, celebrating the god Saturn, was a major celebration extending from the winter solstice on December 21st until January 1st. Furthermore, December 25th was celebrated as the birth date of the Persian god Mithra who was widely worshipped in the Middle East and Mediterranean areas and was popular with the Roman legionnaires. (Santino 1994, p. 178)

Throughout the northern hemisphere people have celebrated and still celebrate with a festival of light at this time. Some peoples saw the winter solstice as the birth of the sun because the days get longer after the solstice. (Santino 1994, p. 183) The birth of Mithra was called the Birth of the Unconquerable Sun. (p. 178) England and the Scandinavian countries burn the Yule log—the mid-winter fire, the mid-winter light. Santino notes, "*Yule* may come from the old Germanic word *Iol*...which means 'a turning wheel' and might refer to the turning sun after the winter solstice." Or it may be from the Old English word *geol*, meaning "feast." The entire month of December was known as "*geola,*" or feast month; a great solstice celebration before Christianity arrived. (p. 181) The menorah is the central symbol of the Jewish Hanukkah, whose 8 candles mark a period of miraculous light when a lamp burned without fuel for 8 days. (p. 182) The Romans made prominent use of candles in their rituals and festivals which influenced Christian symbolism. Christianity associated the birth of Christ ("the Way, the Light, the Truth") with the solstice celebrations as a deliberate means of diverting attention from the pagan celebrations. (p. 178, 183) Swedes celebrate Saint Lucy's Day on December 13 which was originally the winter solstice under the old Julian calendar. They are served breakfast by a young girl wearing a crown of 9 candles and dressed in a white robe with a red sash. (p. 198, 200)

The Solstice, focused on the light of the sun, is the most abstract and spiritual expression of the rebirth motif. The worst, most trying days of winter follow the solstice in Northern environments. It will be two to three months before the earth shows physical signs of coming back to life. Solstice celebrations are about spiritual rebirth and being sustained by the spirit through dark and difficult times until the product of inspiration manifests in a physical manner in spring.

The sense of the future being started at the solstice was represented by the 12 days between the Solstice and the New Year representing the 12 months of the New Year. Rome had a twelve-day Saturnalia celebration and the ancient Babylonians a twelve-day festival of Zagmuh. (Santino 1994, p. 47) England celebrated the twelve days of Christmas in the 18th century (p. 178); my English wife insists our Christmas tree stay up no longer or no less than 12 days after Christmas (until January 6, Epiphany or Twelfth Night).

Foliage and evergreens are the primary decorations throughout the winter season. The holly, ivy and mistletoe used as decorations during the Christmas holidays were believed by the northern Europeans to have special power. The druid priests of the ancient Celts felt mistletoe was so sacred that if "enemies...met each other under it in the forest they were required to lay down their arms, exchange friendly greetings, and keep a truce until the following day." Until recently the English decorated their homes with a kissing bough of mistletoe. (Santino 1994, p. 181)

During the Saturnalia celebration the Romans combined the symbolism of light and evergreens by hanging burning candles from trees. (Santino 1994, p. 180) It was the Germans who brought the custom of the Christmas tree to America. Its roots probably go back to medieval Biblical folk plays involving a decorated fir "tree-of-life" combined with a Germanic custom of putting evergreen boughs and laurels on a pyramid-shaped wooden structure. (p. 182)

The custom of exchanging gifts at Christmas also has its roots in the Roman Saturnalia. In America it is Santa Claus who brings gifts in the night, and his evolution is an interesting one. His name is derived from St. Nicholas, a Turkish bishop of the fourth century who became the patron saint of fisherman because it was believed he could calm storms. He was noted for saving three destitute young women from selling themselves to prostitution by secretly visiting their house at night and tossing a ball of gold through their window. (Santino 1994, p. 192) Many countries celebrate December 6th as his feast day, when he arrives with presents or punishment. When Christianity moved into northern Europe, the St. Nicholas character was blended with Wotan, or Odin, the chief god of Teutonic religions. Wotan was a hairy old man who rode an eight-legged horse throughout the world. His warrior women Valkyries "rode the winds to gather the souls of heroes who had fallen in battle." (p. 194) Wotan and St. Nicholas were both travelers, inspecting and judging mankind's deeds during the darkest part of the year when the future was being determined. Both could raise or lower storms. (p. 196)

The blending of elements produced in Northern Europe a St. Nicholas who rides a white horse, is a hairy old man, and may be accompanied by a black, scary companion who carries bad children off in a sack. (Santino 1994, p. 195) Our American Santa Claus has the bishop's name and robe, rides in a sleigh pulled by the reindeer of northern Europe, and lives at the North Pole. (p. 195, 196) The only hint of judgment is the threat of a lump of coal being left in the stocking of a misbehaving child. (p. 197)

The Reformation English threw out St. Nicholas and substituted Father Christmas, a winter deity with white hair and beard and a crown of holly. He became a gift-giver when the Father Christmas of the English Americans blended with the American German's and Hollander's gift-giving St. Nicholas. Clement Moore's poem in 1823, "The Night Before Christmas," either created or reported on a tradition of a Santa as a jolly old elf with eight reindeer who comes with presents on Christmas Eve. Santa is the gift bringer in America, while either St. Nicholas or Christ is the predominant gift bringer in Europe, either on December 6th or Christmas Eve or day. (Santino 1994, p. 197, 198)

The sun's movement at the solstice towards greater light after its darkest moment is symbolically represented by St. Nicholas's golden gift given in the dark and Santa's gifts delivered in the dead of night. Psychologically this represents the spontaneous arising of a new light of consciousness, a birth or a re-birth out of the dark depths of the unconscious. Winter symbolically is a time of purification by the cold, cleansing winds of the north. The old forms are frozen out, killed off. The unconscious often produces an image of wholeness and healing during its darkest hour. Jung believed *the* main human archetype was a longing for the light of consciousness and an escape from the darkness of unconsciousness. After old forms crumble, a spark, a seed, an idea, a circle of light emerges out of the blackness of yin in the Chinese yin-yang symbol. One becomes fully conscious of how bad the present situation is and painfully aware that things could be better. Judgments have to be made about past actions so old ways can be mended to allow a new and better future to emerge. Casting the fortunes of the future with the New Year on the way is reflected in our New Year's resolutions.

The soul experiences the long, dreary, trying winter following the solstice, while sunlight increases, as a metaphor for gradually increasing consciousness and understanding as one works through one's complexes while being encouraged and guided by a growing sense of an inner spirit. As with the violently changing weather of late winter and early spring, the establishment of a new way of being is never a straightforward path or without what feels like the threat of death and regression into the old, dark forms.

The New Year's Baby, symbol of rebirth and the New Year, is contrasted with Father Time, the old year, and with Saturn of the Roman Saturnalia. Saturn is associated with form, structure and age; births and babies are associated with potential and infinite possibilities. New Year's resolutions are our way of moving beyond old ways of the past and doing something different with the potential of a new year. (Santino 1994, p. 208) January, named after the Roman god Janus, is a god with two faces: one to look into the past, the other into the future.

The Christ child is a solemn image of rebirth and renewal, while the New Year's baby has a sense of playfulness about it. (Santino 1994, p. 207) Santino notes, "Christmas focuses on the future in a solemn and sacred way, New Year's does it with abandon." (p. 205) Transition points, such as going from the past to the future, can be dangerous and are often accompanied by magic beliefs and ritual or spontaneous, licentious, boisterous celebrations. The birth of the New Year is celebrated in a sacrilegious way on New Year's Eve by flouting convention, by drunkenness and craziness, and a form of social leveling by kissing whomever is standing next to one at the stroke of midnight. (p. 208, 209) Flouting social convention and social inversion was also seen in the twelve days of post-solstice celebration in the ancient Persian Sacaea and the Roman Saturnalia festivals when a slave was chosen to be head of the household. (p. 47) Whereas Thanksgiving focuses on the elder energy of one's roots, and Christmas focuses on the children in a family setting, New Year's Eve is celebrated with one's friends and peers, out in the world which symbolizes moving outward from one's roots. (p. 209)

The Super Bowl, coming close enough to the New Year celebration, is associated with the slaying of the old King and installation of the new. Anthropologist Clifford Geertz pointed out, "our rituals, spectacles and sporting contests such as the Super Bowl are symbolic events that help us make meaning out of life." (Geertz 1971 referenced in Santino 1994, p. 51) The meanings derived from the Super Bowl are associated with American male power games and the essence of capitalism. Highly competitive, aggressive team actions performed by specialized, powerful men are played with rules whose enforcement is grumbled about. The playoffs and championships represent climbing the corporate ladder. Financial stakes are high for players' salaries, franchises, TV commercials and betting. The pre-game hype and elaborate half-time shows heighten the sense of contest. This contemporary ritual sets forth the ideals of renewed masculine leadership at the beginning of the year. (Santino 1994, p. 50-55)

Like many sports, football provides a time grid for the year. The beginning of practice in mid-July is a reminder that summer is half over, the first regular season games mark the end of summer relaxation and the beginning of intense fall activity, and the playoffs culminate in an undeclared holiday on Super Bowl Sunday close to the beginning of the New Year. (Santino 1994, p. 53)

Martin Luther King Day in mid-January is the only American holiday that honors a spokesperson for minorities, the outcast and the downtrodden—an American version of the challenge to the social order that occurs at the New Year. The new hope and inspiration symbolized by the New Year's baby is not without considerable threats to its survival from the dominant forces in society or in one's unconscious. The Biblical story of this archetypal dynamic is the Holy Family fleeing to Egypt as Herod killed all male children under two years of age.

Groundhog Day on February 2nd is a day when we focus on the rhythms of the earth reflected in the seasonal activity of its creatures. (Santino 1994, p. 58) It is an unusual little event that attracts more interest than one would expect as it marks the first time we direct formal attention towards the coming spring. (p. 56, 57) If the groundhog sees his shadow when he emerges from his winter hibernation, it frightens him back into his hole and we'll have six more weeks of winter. If he doesn't see his shadow, he will stay out and we'll have an early spring. On Groundhog Day, German and French farmers formerly noted the stirring of the bear in its den as a sign of spring. (p. 58) In old carnivals and masquerades at this time of year people often dressed as the Candlemas bear whose appearance suggested the coming of spring. (p. 59) As bears became scarce, badgers were used as weather prognosticators in Germany. German settlers brought the tradition to Pennsylvania and gave the woodchuck, or groundhog, the predictive honor. (p. 58)

February 2nd is actually a bit too early for hibernating animal activity but the date had been moved up 11 days with the advent of the Gregorian calendar. (Santino 1994, p. 58) February 1 is Saint Bridget's Day in Ireland when the hedgehog looks for its shadow. February 1st was also one of the major Celtic festivals, Imbolc, "in the belly," considered to be the first day of spring. (p. 59)

President's Day, observed as a Monday holiday in mid-February, serves to remember and honor archetypal father energy in America. Washington and Lincoln are associated with strong, wise leadership through the chaos of wars and the breakup of old forms. Their seminal influence helped establish and maintain the American principles of freedom and a unified, democratic government. They represent yang energy as inspiration and basic, guiding principles; yang as abstract founding, unifying and sustaining ideals that hopefully will emerge out of death and chaos.

The major seasonal changes in autumn and spring are in the domain of Hermes as the god of transitions. In the autumn he is associated with the disruptive chaos resulting from the death of old forms and the social inversion of Halloween tricks. Santino further links prankster activities with male aggressiveness and outdoor settings. (Santino 1994, p. 73) Relevant myths include the rape of Persephone and Hermes' relationship to male initiation of the feminine into Hades (see volume 3 of *The Dairy Farmer's Guide*). The situation is reversed in Hermes' winter-to-spring transition that he ruled over as the progenitor of new forms as a result of him being associated with phallic energy, Eros and connectedness. These traits are reflected in our Valentine's Day emphasis on sex, romance and relationship leading to the marital bond. The fertility associated with Valentine's Day and Saint Patrick's "greening" symbolically can represent the first manifestation of principle (associated with the Presidents as archetypal yang energy) into form and physical expression (from **Spring** to **Growth** in the *I Ching*).

Valentine's Day in mid-February is the final winter holiday of the season. Its message is that life is coming soon, spring may bloom from the melting winter snow. (Santino 1994, p. 78) Mid-February is associated with fertility in many folk beliefs. (p. 59) In the Middle Ages it was believed birds chose their mates on February 14th and bird fertility is indeed a sign of spring. Valentine cards often depict birds, turtledoves and lovebirds. Our cupid, derived from the Roman version of the Greek god Eros, is a desexualized and non-threatening rebirth symbol as are the Christmas and New Year's babies. Western Europeans and Americans translate sexuality into romantic love; Pan and other gods of sexual activity and passion are too sensual and bedeviling for our Puritanical tastes. (p. 68) (n 10)

In 469 AD Pope Gelasius chose February 14th to remember the Christian martyr Valentine. One story is that Valentine was beheaded for secretly conducting weddings, violating Roman Emperor Claudius's ban on weddings, which were keeping men from fighting wars. (Santino 1994, p. 68-70) The Church chose to honor St. Valentine in mid-February and set the Feast of the Purification of the Virgin Mary, or Candlemas, on February 2nd in an attempt to siphon off interest in the Lupercalia, a purification and fertility festival held in late winter in the Roman Empire. (p. 60, 68)

Santino writes, "The Lupercalia was a torchlight parade of purification, and it may be where the church's custom of blessing the candles used in its liturgical services arose [as done on Candlemas Day]." (Santino 1994, p. 60) The Lupercalia extended the archetypal theme of birth to the mythical birth of the Roman Empire. This major festival honored the wolf (*lupus* in Latin) who mothered the twin founders of Rome, Remus and Romulus, for whom Rome was named. One of the Lupercalia rites included two boys going to the cave in one of the seven hills of Rome where Romulus and Remus were believed to have been nursed. The boys were anointed with the blood of sacrificed goats and a dog, and then ran through the streets striking people, especially women, with thongs made of goatskin (called *februa*) to promote fertility in the women. (p. 78) *Februa* is Latin for "purify" and the root word of February, the second month in the Roman calendar. (p. 60) Goats symbolize potent sexual drive and they mated in February. (p. 78) Many animals and nature gods were associated with the Lupercalia over the centuries it was celebrated, including the wolf, a hairy wolf-like god called Lupercus, and a bestial Romanized version of the Greek god Pan called Faunus. The festival was conducted largely as a fertility rite for agricultural and pastoral pursuits and to protect the fields and herds from wolves. It was a festival of increase like the Greek festivals to Pan, the goat-god. Festival goers drank excessively in the streets. (p. 60) These festivals held in anticipation of spring often evoked sexuality as part of the rebirth of nature. (p. 70)

A remnant of this fertility ritual survives in the modern day Czech Republic as reported on the front page of the March 28, 2002 *Wall Street Journal*. The day after Easter men and boys whack female relatives and friends across the legs with willow sticks. This is part of a pagan tradition that some Czechs believe bestows lasting youth upon women:

> In some villages, especially in southern Moravia, the festivities get under way as much as a week before Easter with various folklore events that include parading through the streets with animal masks...
>
> After mass on Easter Sunday, boys and men start to prepare for the big event by braiding live willow branches with pieces of ribbon into makeshift switches. By dawn on Easter Monday, a traditional holiday across Europe, many are already hammering on the doors or prowling the streets looking for stray [women] victims.

> The rituals that follow originated at least as far back as the Middle Ages, when boys would use their sticks to lift girls' shirts in a symbolic gesture supposed to enhance fertility. This later evolved into playful swiping across girls' legs and dunking them in water or perfume to cast off impurities and prevent premature aging. As a sign of gratitude, boys are given decorated eggs and men get shots of alcohol. After midday, the tables are turned, and girls are allowed to fight back by throwing water at the men.

In the Lupercalia boys drew a girl's name from a box and then escorted that girl to the festival. (Santino 1994, p. 61) We all have memories of the Valentine box and exchanging Valentine cards in grade school replete with images of birds and flowers associated with sex and romance. The common Valentine image of a heart pierced by an arrow hints of male and female sexual union as well as the risks of making oneself vulnerable by exposing one's feelings. The colors associated with the day, red and white, extend this idea, with red symbolizing the blood as life force, passion and the Devil and white being purity, the sustenance of mother's milk and spirituality. Pink is the combination of these two opposites to produce new life. (p. 77, 78)

Saint Patrick's Day on March 17 is close to the first day of spring on the Gregorian calendar (March 21). Saint Patrick, who brought Christianity to the Celtic Irish, allegedly lit a paschal bonfire to celebrate Easter across a valley from druidic bonfires. This was done to symbolize the Christian faith and re-define the traditional Celtic symbols and customs within a Christian context. (Santino 1994, p. 80) The green and fertility of springtime is reflected on Saint Patrick's Day by the shamrock, the "wearing of the green," leprechauns and green beer. The druids associated the shamrock with the regenerative powers of nature. Saint Patrick used the shamrock to convey the doctrine of the trinity (three-in-one) while the Celtic-style cross represents the cross-like shape of the shamrock. The green leprechauns were believed to live in the hills and valleys and represent the spirits in nature and the fertility of the earth. (p. 80-82)

Saint Patrick also brought the art of distilling spirits, symbolic of a cultural form as something derived or refined from a natural form. In the US, Saint Patrick's day is associated with much public drinking. Licentiousness and inverting rules usually accompany large-scale public celebrations of increase and fertility. (Santino 1994, p. 82) Everyone wants to be Irish on Saint Patrick's day, a type of social leveling. The greenery, drinking and good feelings associated with the day contribute to the sense of coming to life at the beginning of spring. (p. 84)

The February masquerades and spring carnivals of ancient and medieval times survive to the present day as Mardi Gras, Carnival and Fastnacht (Germany and Switzerland). The period from January 6 (Epiphany, Three Kings Day, or Twelfth Night) to Ash Wednesday is called Carnival, which "bridges the gap between Christmas and Easter, winter and spring." (Santino 1994, p. 86, 94) The Carnival culminates in Mardi Gras, French for "Fat Tuesday," the day before Ash Wednesday which is the beginning of Lent. "Fat Tuesday" possibly refers to an old practice of slaughtering a fattened ox for a final meal just before Lent or frying foods before Ash Wednesday to use up the animal fat in the kitchen. (p. 89) Lent means "lengthening days," and is 46 days before Easter. (p. 86, 102) It is a time of personal sacrifice with fasting and abstinence in remembrance of Christ's sacrifice on the cross. (p. 103)

Carnival is celebrated in Europe, the Caribbean, Central and South America—and New Orleans. (Santino 1994, p. 86) It begins in New Orleans with the King's Ball on the Twelfth Night, with balls continuing on a weekly or bi-weekly basis until Mardi Gras. Carnival may be a corruption of the Latin *carne vale,* which means a farewell to meat. (p. 88) It probably means a farewell to flesh and fleshy delights given the wild and fleshy nature of the celebrations. During Carnival in New Orleans, women may be found in various states of undress, garters and underwear are tossed from floats in parades, and the famous transvestite parade marches through the French Quarter. Street parties and parades occur throughout Carnival season, with elaborate and original floats and spectacular costumes. (p. 89, 93)

The French brought the Parisian Carnival tradition with them when they established New Orleans in 1718. (Santino 1994, p. 89) Over the years it incorporated many other cultural influences including Anglo-Saxon private clubs (krewes) that conducted parades. (p. 91, 92) The krewe-sponsored parades each have their own king (p. 90), reminiscent of French medieval carnivals where people formed pretend kingdoms for celebrations. (p. 92) Rex, The King of the Rex parade, is "the King of Tomfools, Lord of Misrule, and King of all Mardi Gras." (p. 94)

Mardi Gras epitomizes the principle of social inversion with people from all classes and ethnicities engaged in raucous celebrations that mock and challenge social pretensions. (Santino 1994, p. 92) It is correctly described as a modern descendent of the Roman Lupercalia.

The church was unsuccessful in its attempts to suppress the Roman carnival after the Apostle to the Germans, St. Boniface, complained in 742 AD of the "lurid carnivals" of the Germanic Franks and Alemannia. (p. 90) Masked, rowdy and costumed revelers in central and south Europe flocked through medieval streets, ridiculing the clergy and demanding gifts of food and drink from the wealthy. (p. 48)

An interesting carnival emphasizing social inversion is The Feast of Fools celebrated in France and England which satirized the church and her powerful authorities. In England the feast was celebrated on December 28 and known as the Feast of the Boy Bishop. The lesser clergy in France, usually poor peasants, took the roles of the Lord of Misrule and the Mock King. The mass was parodied by bringing donkeys into the church, men dressing as women, shoe leather being burned instead of incense, and a beggar's banquet celebrated on the altar. (Santino 1994, p. 96) The church sponsored the feast to release anti-clerical sentiment and as an enactment of the Biblical statements to exalt the humble. The rowdiness and hostility that often erupted and the fear of undermining authority led to a church ban in 1431. The festival survived for another 100 years with elements still seen in the revelry of New Year's Eve and Halloween pranking. It was especially popular in medieval France, where April Fool's Day, which echoes its name, is still widely enjoyed. (p. 96, 97)

Some historians believe the jokes, fooling and sending friends on fool's errands on April 1, April Fool's Day (our only remnant of a once major carnival), have to do with the change to the Gregorian Calendar. The French adoption of this calendar in 1564 switched New Year's from March 25 to January 1. The old new year's festival, The Feast of Fools, extended from March 25 to April 1 in certain areas of France and culminated in gift giving. Those who kept the old calendar were sent mock gifts of derision on April 1 and called April fools or April fish. The fish remains the primary symbol of this day in modern France, where chocolate candies in the shape of fish can be found everywhere on April 1. (Santino 1994, p. 97)

The pre-Christian fertility celebrations associated with the spring or vernal equinox get re-enacted by students on the beaches of Florida during the infamous Spring Break, occasionally erupting into riots. (Santino 1994, p. 101)

The Spring equinox influences the unusual changing date of Easter which is the first Sunday after the first full moon following the spring equinox. Easter celebrations participate in spring's theme of fertility and rebirth. In the midnight mass, the Pope wears a white robe and lights a large white candle of the "new-fire" to symbolize the hope and new light of the resurrected Christ. (Santino 1994, p. 103) There is the unusual association of eggs and Easter bunnies with Easter. Throughout the world eggs are ancient symbols of birth, new life and renewal. (p. 110) The Chinese traditionally give eggs to celebrate the birth of a boy. In Eastern European and Slavic countries, eggs were symbols of the sun long before the coming of Christianity, with ancient traditions of elaborately painted eggs linked with spring rituals. Rabbits are ancient symbols of fertility and sexuality and Easter lilies and "new Easter clothes" extend the theme of springtime, the rebirth of the earth, and a renewed psychological life. (p. 110, 111) (n 11)

Christ's Last Supper is believed to have been a Jewish Passover seder. (Santino 1994, p. 104) Passover is a 7 or 8 day feast beginning on the 15th day of the month of Nisan according to the Jewish lunar calendar. This calendar is based on the moon's revolving around the earth rather than the earth revolving around the sun. (p. 105, 106) Passover ritually relives a series of events in Jewish history where God delivered the Jews from Egyptian slavery and returned them with Moses' leadership to the Promised Land of Israel. Deliverance began with the Angel of Death killing the first born of every Egyptian household. The Jews were told to smear their door frames with the blood of a slaughtered lamb as a sign to the Angel to pass over their household. Biblical evidence seems to indicate that the Hebrews had already celebrated at this time an ancient and now-forgotten spring festival involving lamb sacrifices. The modern Passover holiday is a happy, sacred celebration around the theme of freedom. Much wine is drunk during the seder leaving the people in good spirits. (Santino 1994, p. 106) Passover, Easter and spring are psychologically associated with the release from death and the bondage of old forms and complexes, symbolically associated with winter, into the rebirth and freedom and exuberance of new life.

From May through September in America we lack the special days of significance and power *specifically* associated with the compelling symbolism and traditional imagery of seasonal, solar and agricultural changes. Most of the days—Memorial Day, Flag Day, Independence Day, Labor Day, Mother's and Father's Day—could be held at any time of the year. These holidays aren't derived from the seasons but grafted onto them and related to in a seasonal way, such as Memorial Day being considered to be the beginning of summer and Labor Day its closure. (Santino 1994, p. 124)

Summer is dominated by the sun with sunlight to be enjoyed in an active manner. We try to extend the light with daylight savings time. Life is lived outdoors as much as possible with outdoor music festivals, parades, beach vacations, backyard barbecues and camping trips. There are many ethnic and religious celebrations, local and regional activities, and state and country fairs. (Santino 1994, p. 124, 135-139)

May transitions us from spring to summer. The May flowers of spring, brought about by April showers, are celebrated on May 1st. May Day is a major holiday in many parts of the world but not in America. The Puritans in America extended the Protestant reforms begun by Cromwell in England when he banned May Day celebrations in 1644. Cromwell suppressed traditional holidays and calendar customs as forms of ancient idolatry. Restoration Great Britain abolished the anti-festival laws of the Cromwell era, but the American May Day never recovered from the unflagging persecution of the Puritans. (Santino 1994, p. 112, 113)

May Day is a descendent of the Roman Floralia festival which honored the goddess Flora. One Roman activity included young girls winding flowers around the columns of the temples. (Santino 1994, p. 113) Floralia fused with the Celtic celebration of Beltane celebrated on May 1st, a quarter day in the solar year between the equinox and solstice. Beltane was celebrated with bonfires, and fairies were believed to be very active. From the Middle Ages into 19th century Europe, May Day was celebrated with such activities as women washing themselves in the dew to preserve youthfulness. Flowers were gathered in the woods and fields for May baskets and decorations. A tree and later a permanent maypole as a town landmark were decorated with flowers and herbs hung with colorful streamers. People feasted and danced around the May pole. (p. 113, 114) The phallic May pole piercing a circular wreath of flowers symbolized the sacred union of the male green god of vegetation with the earth goddess in her form as the goddess of flowers. Many villages chose a May King and Queen to embody the goddess and her consort, which often included their mating to insure fertility of crops and animals.

May is named after the Greek goddess Maia, goddess of the Earth and Hermes' mother. American Catholics celebrate the feast day of Mary in May, designated the month of Mary. This can include the choice of a young girl as Queen of the May and placing a crown of flowers on the statue of Mary. (Santino 1994, p. 117)

All American summer holidays are civil. Flowers associated with summer are joined with the flag to celebrate our national holidays outdoors with picnics, parades and barbecues. Memorial Day, the last Monday of May, is considered to be the unofficial opening of summer. It began as a memorial for Civil War veterans. Today we recognize the armed forces while many people remember specific deaths of family or friends. (Santino 1994, p. 118-120)

Graduations, which mark both an ending and a beginning, occur at the end of May or first part of June. June is also the traditional month of weddings. Both rites of passage are intrinsically fitting ceremonies to open the summer season (Santino 1994, p. 122, 123), symbolically associated with active growth and filling out the new forms which were seeded in the spring.

Baseball provides a seasonal sports grid carrying us from the optimism of spring training in late winter to the mid-summer All Star game in July to the end-of-summer climax with the World Series in mid-October. (Santino 1994, p. 123) Outdoor seasonal jobs, agriculture and tourism add a summer dimension to the work force. (p. 124)

July 4th serves the useful function of being *the* major summer festival, occurring at the time in the solar year when we begin to feel the full force of the sun's energy. Independence Day is an example of how the history of a society changes the date and nature of a traditional celebration, in this case Midsummer Day on June 24th. (Santino 1994, p. 131)

Midsummer Day, like Christmas, was based on a solstice and occurred a few days after it. Midsummer evenly divided the season from May 1, the end of Spring, to August 1, the beginning of harvest and the Lughnasa festival in Celtic tradition. (Santino 1994, p. 131, 132) Spirits and fairies were believed to roam on Midsummer eve as presented in Shakespeare's *A Midsummer Night's Dream.* (p. 132) The longest day of the year had traditionally been associated with magic. (p. 128) Midsummer Day was celebrated by bonfires on the preceding eve, burned as part of magic rituals to mimic the sun in its glory and to give it strength as the days grew shorter. A custom of jumping over fires supposedly parallels the sun crossing the midpoint of the solar year. The bonfire tradition continues today among the Swedes, Finns and Lithuanians and often in conjunction with Independence Day celebrations of other nations. The Catholic Church tried to appropriate the pagan energies by renaming

Midsummer Day as St. John's Day, after John the Baptist. St. John's Eve then became the night of fairies and bonfires throughout Europe. It is a holiday of rejuvenation symbolized by Christ's baptism, symbolically related to the bonfires magically rejuvenating the sun. (p. 131, 132)

The Celtic harvest season began on August 1, one of the major festival days (Lughnasa) in the old Celtic calendar. (Santino 1994, p. 148, 164) The growth cycle of the land begins to decline after mid-August, and the summer heat and sunshine wanes through August and September as fields fade and the harvest ripens. (p. 141) The drama of the fall season is about to begin.

September is a transitional month like May. The placement of Labor Day was chosen to fill a gap in legal holidays between July 4th and Thanksgiving. It is the traditional end of the summer vacation period and children head back to school—a rite of passage. Football season begins, the September stretch run is on in baseball, and presidential campaigns begin in earnest during election years. (Santino 1994, p. 140)

It begins to feel like fall in the northern half of America around the autumnal equinox on September 21, the first day of fall. Jews traditionally begin the year near the autumnal equinox, and their new year festivals celebrate endings and beginnings as do the new year festivals of all peoples. Jews are judged on Rosh Hashanah, New Year's Day, the first of a solemn ten Days of Penitence when the gates of heaven are open to retract vows, right wrongs, and make amends to God and other people. Fates are inscribed by God in the sacred Book of Life and sealed for the coming year on the holiest day of the year, Yom Kippur, The Day of Atonement, after which the gates of heaven are closed. Sukkoth, a harvest festival established by Moses, is celebrated a few days after Yom Kippur. For eight days all meals are eaten in a sukkah, a booth or tabernacle built in the backyard without using nails and with a roof built of leaves and twigs. (Santino 1994, p. 143-145)

And so completes the cycle of the year, marked off by days and months. Even their names have meaning. Our months have Roman names. January is named after Janus, the Roman deity with two faces and February is derived from the Latin word *februare*, "to purify." March is named after the war god Mars, and indeed March weather often feels like a war between the forces of the cold dead of winter and the warmth of spring and life, establishing the Mars-Aphrodite link. April comes from a Latin word meaning "to open," referring to the time of year when the earth opens to life. May is named after the Earth mother goddess Maia, while June is named after the Roman goddess Juno, wife of Jupiter. The other month's names have no seasonal meaning: July is named after Julius Caesar, August after Augustus Caesar, and September, October, November and December simply refer to the number of the months in the old Roman calendar; the 7th , 8th, 9th and 10th months respectively. (Santino 1994, p. 19, 20)

The days of the week are derived from Germanic names except for Saturday, named after the Roman god Saturn. Sunday is named for the sun (*Sonntag,* sun-day); Monday for the moon (*Montag,* moon-day); Tuesday for Tyr, the Norse god of war; Wednesday for Wotan (Odin), the chief god of the Norse pantheon; Thursday for Thor, the Norse god of thunder; and Friday for Frig, the wife of Wotan. (Santino 1994, p. 20)

Days turn into weeks, weeks into months, months into seasons. Life cycles on. "In paying so much attention to the nuances of a landscape going through the changing seasons, the painter is in reality expressing the state of his own soul." We move towards wholeness and integration in cycles aided by a consciousness and attunement to the cycles of the seasons in nature.

CHAPTER 3

Planet of the Insect

Entomology was my last 4-H Club project and it was the most interesting. I pursued the study of insects through to getting a Ph.D. in entomology at Berkeley. My thesis, "Factors Influencing the Susceptibility of the Beet Armyworm, *Spodoptera exigua* Hubner, to a Nucleopolyhedrosis Virus," proved there were age-significant differences in susceptibility to the virus. It demonstrated that I knew a lot about very little. My psyche was backed into a corner; my boyhood connection with nature was lost as I lived an arid scientific life where nothing was real until statistically proven to be true.

My problem is not only personal, it is a basic problem in Western culture. The dark, earthy, feminine, symbolic and sensual are split off from the rational, scientific and intellectual: Apollo and Hermes behave more like enemy combatants than friendly brothers. This can be changed. We can be guided by a myth from our cultural roots—The Homeric Hymn to Hermes (see volume 3 of *The Dairy Farmer's Guide*)—to restore balance to the Western soul and establish a spiritual and psychological base and framework for our lives and for environmental action. Basic changes are needed in our worldview and in our educational systems.

A surprisingly good place to start is with our relationship with insects. It is much like our relationship with the unconscious. We are surrounded by an immense number and variety of insects but we are barely aware of them just as we are immersed in our psyches in a rich unconscious background of immense and unrecognized potential. Only a tiny fraction of insects negatively impact humans, but these often draw out a heroic, us-against-them extermination approach. With the unconscious, it is largely through our pathologies that we become aware of its existence. The fearful response and heroic stance we take against the unconscious forecloses our connection with the wellspring of our energy and potential source of beauty and rich complexity in life.

Jung couldn't stand insects, considering them to be alien, extra-human creatures fit only to be skewered on pins for scientific study. (Jung 1961a, p. 67) This is one of the rare instances when Jung was way off the ecopsychological mark. I write this chapter as an entomologist and a Jungian to redeem insects in the eyes of their human beholders, recognizing their significant place amongst life forms and their deep impact on the human psyche. A holistic exploration of a life form illustrates how we can connect to the environment by appreciating what is around us and how we can begin to develop a sense of the sacred. This is done by using dreams, myths, stories, scientific information, Hillman's imaginal psychology and examples of synchronicity. We will begin by looking at the psychological and spiritual dimensions of human connections with animals, particularly the significance of animals in dreams, then look at scientific information about insects that intrigues the psyche and draws us closer to nature. Finally, we will look specifically at the psychological and mythical dimensions of insects and our relationship to them which includes synchronicities involving insects. Synchronicity demonstrates that the indigenous worldview, "we are all related," is true at *the* most fundamental level.

I. Our Relationship with Animals

Indigenous peoples are exemplars in their relationships with animals. Brave Buffalo, a Teton Sioux, wrote:

> Let a man decide upon his favorite animal and make a study of it, learning its innocent ways. Let him learn to understand its sounds and motions. The animals want to communicate with man...Man must do the greater part in securing an understanding. (Densmore 1918, p. 172)

Every species deserves our respect and admiration as an emanation from the known and unknown forces in the universe, forces we can contact by communion with a particular species. Animals can teach us how to be human, vulnerable, and whole with all that is. (Sams and Carson 1988, p. 13) The essence, the spirit of the animal, can help a person in need of specific talents and in understanding their role in the mystery of life. (p. 13, 14) Many indigenous cultures believe the more simply constituted organisms like insects were a deliberate design to enable them to serve as better guides and messengers of creative powers.

(Lauck 2002, p. 27) They believe there are no chance encounters with organisms in a world they regard as sacred space; each meeting is intentional and purposeful. (p. 29) One can receive a revelation from any creature because it is an aspect of the whole. (p. 67) Indigenous peoples call upon animals for power and healing, whereby one asks to be drawn into harmony with the strength of the creature's essence. Healers ask their spirit animals to enter them and convey healing energies.

Species are believed to transmit their qualities to us through such acts as eating them, being near them, or even by being bitten or stung. (Lauck 2002, p. 94, 95) The latter is interpreted as a warning, a call to action, or a transmission of power, not as a random act or judgment: the person was to be alerted and called to reflect. (p. 160) Directly confronting the pain and distress of a bite or sting becomes a vehicle for attaining wisdom in Buddhism and in shamanism. (p. 32)

Every individual feels a connection with certain animals that can be discovered by consciously looking at, as Brave Buffalo suggests, what animal attracts us or by the animals in our numinous dreams. Native American vision quests often reveal a spirit animal for the individual and certain ceremonies aid the process of discovering one's spirit animal. I participated in a shamanism workshop where the leader shook a rattle over my right shoulder, then over my head and down the left shoulder. When the rattle reached the top of my head, I got a clear image of a particular animal. I realized I had long admired that animal and I have honored that connection ever since. Ralph Red Fox, a Cheyenne holy man, said it is not wise to reveal and boast of your spirit animal, rather let its influence manifest in how you act.

Jung felt a deep connection with animals. He found college physiology to be

> thoroughly repellant because of vivisection, which was practiced merely for purpose of demonstration. I could never free myself from the feeling that warm-blooded creatures were akin to us and not just cerebral automata... My compassion for animals...rested on the...foundation of a primitive attitude of mind—on an unconscious identity with animals. (Jung 1961a, p. 101)

Jung talked about the "primitive man within myself—a world which...borders on the life of the animal soul." (Jung 1961a, p. 160) (see volume 1, Appendix B) He said it is difficult to tell a person they should become acquainted with their animal and assimilate their animal (Jung 1976, p. 70):

> We have an entirely wrong idea of the animal; we must not judge from the outside. From the outside you see, perhaps, a pig wallowing in mud...But it is not [dirty] for the pig. You must put yourself inside the pig. The pig is convinced...he is a very nice and law-abiding citizen of the world, whose daily job it is to sniff through the dirt. [*Visions Seminar* I (9 December 1930), Notes of Mary Foote, p. 282]

Jung challenges us to look at the world with an animal eye and imagine being in an animal body. (Jung 1988, p. 71; Hillman 1982, p. 318, 319 note 13) This is one Jungian response to the deep ecologist's call for a deeper connection to the world around us, a connection that will serve as the base for environmental activism and protection. To be able to look at the world through an animal eye requires imagination on our part, a creative process, a process initiated, stimulated and guided by the animal we are studying (Brave Buffalo). Letting the numinous animals in our Big Dreams allure and fascinate our consciousness naturally draws us to a study of the animal. (see the paragraphs on Aphrodite near the end of Appendix K in volume 3) Our animal spirit guides can be like those in the fairy tales and mythologies of the world, leading us on a path with heart and meaning. The imagination must come from the heart (volume 3, Appendix K) to incorporate an experiential dimension; it must be an embodied, corporal imagination. To see the world with animal eyes begins at the level of a particular animal, be it a bird soaring, an ant crawling through the grassy jungle of a backyard, or a mouse moving through an underground tunnel. Our world might be the complex micro-environments of insects with intense sensitivities to light, heat and cold. We would be attracted to different smells and might find live flesh, carcasses or dung to be our favorite dining venues. Being in an animal's body offers an infinite variety of imaginations, be it a bird's body with powerful chest muscles for flight or living within the armored skeletons of insects and other arthropods, skeletons periodically shed because they are outgrown, leaving them tremendously vulnerable during the interim.

Seeing the world through another creature's eyes has the transformative effect of relativizing our position. We develop compassion for another worldview and other motivations. Each species becomes a

Greek god with a complete world gestalt as seen through their perspective. (see volume 3, p. 2, 3 and volume 3, Appendix K)

Indigenous peoples use props to help them become ensouled with an animal spirit. They may wear some part of an animal such as an animal skin robe with head and feet intact: think how much we can change just by putting on a mask or costume at Halloween. Natives produce the sounds of the animal, and do imitative dances so their whole body can experience the unique gaits, postures, play forms and routines. The Lakota Sioux, as do many indigenous cultures, form societies with other members who have dreamt of the same animal species. Visions and Big Dreams often include songs and music conveyed by an animal using humans as its mouthpiece. Ritual activities and ceremonial objects depicted in visions provide the ceremonial base for a dream animal society, with the ceremonies amplifying the experience and connection to the animal. The experience feels like one is an embodiment, a human vehicle that gives voice, music and expression in art, intellectual concepts, and visual forms to the spirit of a particular animal.

I had a powerful experience of what felt like being in another body during a two-day vision quest on the Rosebud Reservation in South Dakota. While I was moving between the north and east flags on a prayer round, I had the fleeting, overwhelming sense that my left leg was that of an animal I feel a particular connection with. It was uncanny how different the shape, feel and movement of my leg felt. This experience made me realize how our white culture is in many dimensions removed from a depth of connection to nature described by indigenous peoples.

Even modern Westerners have dreams where animals speak, behave in unusually friendly ways, or appear in a numinous manner. Jungians respond by reading about the biology and habitat of the animal and visiting it in a zoo. There one can notice its behaviors and smells and how it moves and responds to things. It is good to remind yourself that this *is* what indigenous cultures would call *your* animal spirit. It is even better to visit the animal in its natural habitat: this draws one into a particular ecosystem. You will find yourself collecting pictures and objects associated with that animal, having an animal calendar, etc. Reading fairytales and myths about the animal illustrates how it strikes the psyche at the mythic level and it helps one live a mythic life.

Most people come to recognize more than one spirit animal, but it takes only one to have a significant alteration of one's worldview and establish a deeper connection to nature. Professor Herb Martin describes such an experience in an article in *Crossroads—The Quest for Contemporary Rites of Passage* (1996). Martin is Assistant Professor of Cross-Cultural Competency at California State University at Monterey Bay, specializing in world mythologies. He trains teachers how to use the myths of heroes and heroines of various cultures to facilitate the experience of an inner vision quest in schools. His personal story illustrates the transformative power of a spirit animal in modern day Western youth.

Dr. Martin had a recurring nightmare beginning at age 11 of being chased by wolves. He became "simultaneously frightened, awed, and somehow *inspired*" by them, leading him to become a child expert on wolves. He wrote poems, limericks and term papers on them. When 12 or 13 he discovered a story "that changed his life and gave it meaning." It told of a man on his annual solitary fishing trip to Canada where one year a talking wolf visited his campfire. He said he had been a Medicine Chief, Running Wolf, in his previous life and the fisherman was camping on the tribe's sacred burial grounds. "He spoke of the beauty of the land and many secret wonders of the Universe...As the wolf spoke of all these things in such a sincere and loving way, the Man's sense of peace and respect for this place deepened." The man fell asleep and awoke the next morning with the belief that he had only dreamt of a talking wolf. He was astonished to find an old medicine bag beside the campfire with beadwork depicting a running wolf. (Martin 1996, p. 314, 315)

Herb would sob every time he read this story throughout his teen years, always being inspired by it. It initiated him, as he describes it, into "the Great Mystery that is Life." He adopted the wolf as his totem animal, giving himself the name Running Wolf, and started on a path that eventually led him to become an expert on mythology. (Martin 1996, p. 315)

Professor Martin's story illustrates that one's spirit animal may initially appear in a negative and frightening form. If unconscious energies are ignored they don't simply disappear; they turn more negative until we ignore them at the cost of our psychic well-being. Herb resisted the typical response of fighting negative forces that come to us and thereby discovered the secret of shamanism—"the one who attacks and destroys the initiate is the one who becomes the ally and teacher after the trials of initiation has been endured." (Lauck 2002, p. 256) Goethe wrote:

> And so long as you haven't experienced
> this: to die and so to grow,
> you are only a troubled guest on the dark earth.
> (Goethe 1980, "The Holy Longing," p. 70)

Sandy Klipple's spirit animal was revealed to her in quite a different manner. She had a vision of a mare nursing a colt right in the middle of the living room at a social gathering. Later she had a dream of terrified horses and Native Americans pleading for her to help. The dream changed her life. She felt compelled to do something and became an activist instrumental in stopping a mining project in northern Wisconsin that would have contaminated waters on and off the reservations. Sandy credited the power of the horse for giving her the courage and stamina to stand up to government agencies and big corporations through testimony, contacts, letter writing and organizing. She founded an ecofeminist network to warn about methyl mercury contamination of water supplies caused by coal-powered electricity generating plants in the Midwest.

Our Western cultural history includes an ancient Greek custom of going to Aesclepians, or healing temples, for treatment of illnesses. This often featured a dream incubation period after performing purification rituals. One hoped to dream of the god Aesclepius or an animal associated with him—the cock, dog or snake. The priests of the temple helped interpret the dreams and prescribe healing measures that could include dietary changes and activities. An important discernment was to determine which god or goddess the person was to worship as the central element of their healing. Each god and goddess is a little cosmos, having their particular flowers, birds, groves of trees, type and location of temples, and rituals. By these means one learned how to see the world from their perspective, i.e., one's worldview was changed. The process is similar to undertaking a vision quest or doing an indigenous healing ceremony, with the holy man helping to discern what animal or power in nature one is to "follow." It could be a rock or a power like the wind.

Jung said we should realize that each animal has a specific place and activity within an ecosystem—it is not a chaotic, boundary-less domain. The abilities and forms of animals can inspire and guide us because they fascinate our ego consciousness:

> For the animal shall not be measured by man. In a world older and more complete than ours they move finished

> and complete, gifted with extensions of the senses we have lost or never attained. (Beston 1981, p. 25)

Humans can become co-creators of the universe, an alchemical premise, by developing imaginative responses from the heart to nature's mystery, beauty and complexity. An important human contribution is to give nature an objective existence (Jung 1961a, p. 255, 256) and expressing its inspiration in ways that enhance and grace the mystery. Frank Lloyd Wright described his organic architecture as conveying in architectural form the distilled essence or spirit of the nature of a site. The organic architect helps one see and connect to the immediate environment via the form and composition of the building and how the outside environment is seemingly brought into the living space. One feels inspired, grounded and contained in the architectural space and in viewing the building on the site. Our buildings, public and private, can become vehicles, instruments, to further and deepen our connection to nature.

James Hillman challenges psychology to enter a new domain in its relation to the natural world, a domain more like the indigenous mindset. His approach is two-fold: first he elucidates the elements of our scientific and psychological worldview that limit our connection to the outer world, then he offers an alternative worldview from within our Western cultural tradition that re-ensouls the world.

Subtle psychological limitations occur in relation to animals in what Hillman calls "interiorization." (Hillman 1982, p. 320) This sees animals as symbols, such as the pig being symbolic of the devouring earth mother (p. 324), or animals as part-functions of humans—cunning like a fox or courageous like a lion. (p. 312) Other subjectivations include metaphoric labeling of human traits as being shy "mousey" behavior, "piggish" greed or "swinish" sexuality. (p. 303, 305-319) Animal images are limited if we take them as representing an organ (insects as the external equivalent of the vegetative nervous system) and snakes as a symbol of an instinctual drive or a penis (Freud). Freud saw wild beasts in dreams as representing passionate impulses that scared the dreamer. (p. 312, 313) If dreams are taken as being in compensatory relationship to the conscious mind, there is a tendency to associate animal images in dreams with the "unconscious psychological life, the sins and wishes of emotions which the animals have been forced to represent. " (p. 320) Animals in dreams may also be idealized as "the wisdom of the body as the natural superiority of the million-year-old inner animal." (p. 318)

These approaches may help us feel a deep organic kinship with animals while maintaining human superiority. The animal loses its sense of transcendent otherness, of "its ownership of itself as a self-possessed creature with its own nature not assimilable to mine." (p. 320, 321)

Hillman re-ensouls the animal when he confesses:

> I do admit to taking the animal in the dream unvaryingly as its most significant element because it is the place where the psyche opens into forms, beings of mystery and beauty who are creatures as we are and yet remain 'other.' (Hillman 1982, 304)...The animal in a dream presents not *my* body...but *its* body, the essential otherness of *its* shape and motion, the self-display of *its* physical presence. (p. 319)

The ultimate challenge in working with animal dreams is to study the animal to the point of being able to see the animal with an animal eye as would a shaman. One attempts to grasp the unique otherness and spirit of the animal by an intimate examination of the gestalt of its presentation. "The interior selfness of the animal appears in its displayed image," Hillman writes. (Hillman 1982, p. 324) Each animal is a concrete, unchanging, universal expression of its species revealed in an image of "a complexity of habits, presentation, style" (p. 304):

> To read the animal, to hear it speak, requires an aesthetic and ecological perception...of what is presented...[calling for] an appropriate, appreciative response, grateful that it is even there, that it has come to the dream, and that this visitation is a momentary restoration of Eden...In the dream of images [is] an original co-presence of human and animal...in the Garden from which animals were never ejected. (p. 303)

(For an elaboration of Hillman's intriguing approach, see volume 3, Appendix K)

Responding to the world like an animal is to be immediately, directly engaged in a sensuous relationship to it. *Religio,* after all, means a "paying attention to":

> The Animal Soul is that which perceives and feels, without it he may not perceive or feel the Joys of the Universe. Despised as the Fallen Daughter, it is our greatest Treasure, for it is the Kingdom of Heaven on Earth. (Achad 1973, p. 87)

The animal is partly divine and the divine is partly animal. In indigenous cultures, killing animals on the altars of the Gods "feed[s] the animal in the God" (Hillman 1988, p. 71):

> Gods retain this animal eye. Their animal heads and animal masks display their animal consciousness. The head of the animal on the human torso maintains that lower, immanent vision of creatureliness, creator and creature, God and animal, in the same figure. (Hillman 1982, p. 325)

Jung said it was important for Christianity to regain the theriomorphic (animal) form of God. (Jung 1961a, p. 215, 216) Hillman's idea of the objective psyche is that "images exist in their own right. Like the fox in the forest which is not mine just because I see it, so the fox in the dream is not mine just because I dream it." (Hillman 1982, p. 320) Indigenous cultures retain the sense of the transcendence of animals, not reducible to psychological percepts or instinctual processes. Birds, animals and other natural phenomena are seen as messengers from the Source, the Creator; the animal form of God, god in animal form. In most societies the animals were once gods, not symbolic representations of gods, but the gods themselves. The recognition of otherness is the basis of all religious feeling and may have originated with a religious interpretation of a sacred otherness of animals. (p. 321) (n 1)

A woman dreamt:

> There are a lot of tiny animal creatures who have fragments of an original knowledge, guidance, which they preserve jealously with great tenacity to keep it, and them, alive. I watch them guarding these fragments, and scurrying around, building, or re-building, their living place, and I feel reassured and hopeful. I know (without being able to say why it is so) that these holy fragments will last forever, go on forever, yet will help only if these creatures give them their utmost care and attention. (Hillman 1982, p. 327)

We can still feel the divinity of animals in our dreams:

> There, their souls and ours meet as images...The restoration of the animal kingdom is...a restoration of ourselves to that kingdom via the dream...The dream itself encloses us protectively in the saving ark, in the origination garden, and there in the dream, we may recover the habits of the crab and the mouse, the knowledge of the pig, the animal

> coat, the animal tail, the animal eye. (Hillman 1982, p. 328, 329)

We can strive to develop this dream sense in our waking life by a religious ("paying-attention-to") response to dream imagery and embodying that essence in our lives as an indigenous person would do.

II. The Basics about Insects

Before turning our attention to the divinity of insects, we should know some basic facts about them, learning how scientific knowledge can deepen our sense of awe, mystery and soul-fullness. The most basic fact is that insects are arthropods, one of many phyla (major divisions) in the animal kingdom. Arthropods include the spiders, lobsters and crabs, scorpions, centipedes and millipedes, and the extinct trilobites. Arthropods (fig. 1) have jointed feet ("arthro"-jointed, "poda"-feet), a segmented body, and a skeleton on the outside of the body (exoskeleton) made of a tough material (chitin) that is periodically shed as the animal grows. They have a tube-shaped digestive tract (alimentary canal) with a circulatory system usually consisting of one vessel open at both ends and located above the digestive track. (fig. 2) The blood moves through a body cavity and not in vessels. The nervous system consists of a ganglion (its brain) in the front above the alimentary canal and a ganglionated nerve cord below the alimentary canal. Respiration (breathing) is by gills, trachea (tubes) and spiracles (holes in the exoskeleton).

The class (group) Insecta is distinguished from other arthropod classes by having three body segments: a head for feeding and sensation, a thorax for locomotion, and an abdomen for digestion, excretion and reproduction. Other distinctions are one pair of antennae, three pairs of legs (spiders have four pairs so are not insects), and often one or two pairs of wings. (figs. 1 and 2)

Insects are so ancient that they evolved together with the land plants. The first land plants formed low ground cover about 400 million years ago. Insects emerged about 50 million years later having evolved from the annelids—the worms. By the end of the Carboniferous period 300 million years ago, plants had produced the great coal forests; the amphibians (frogs, salamanders, etc.) and their immediate descendants, the reptiles, were flourishing; and cockroaches and dragonflies were similar to today. Insects had completed the most crucial events in their evolution and the prototypes for many orders of insects were established. By 200 million years ago the diversity of insects was similar to what it is today, along with many extinct groups. (Eisner and Wilson 1977, p. 3, 4)

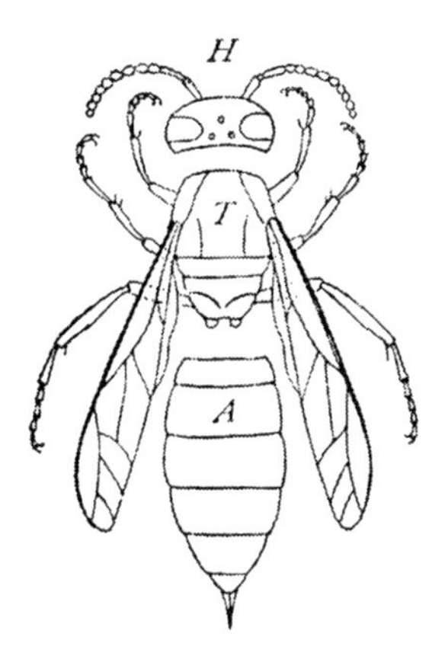

Figure 1. Wasp with head, thorax and abdomen separated. (Comstock 1950, p. 27)

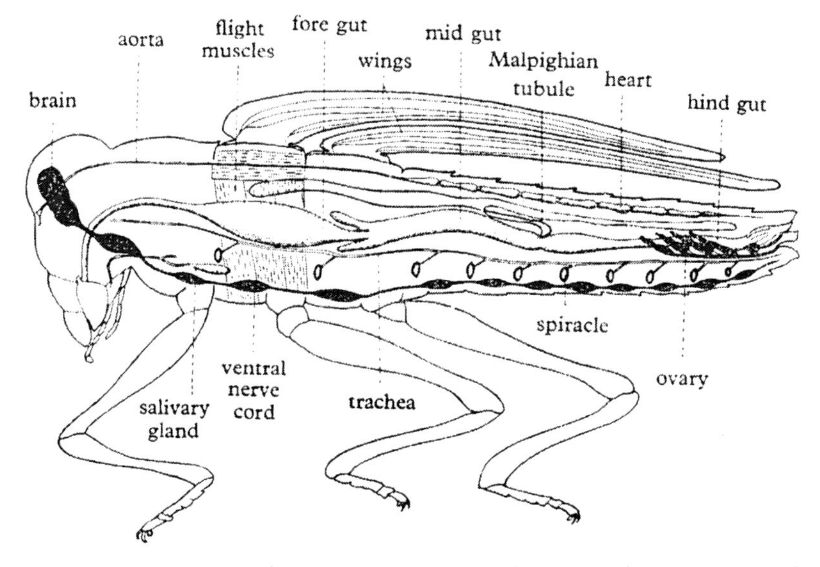

Figure 2. Diagram of an insect to show the typical structure and features of arthropods. (Borradaile et al. 1961, p. 429)

Insects are the conquerors of the land, all but owning it in an ecological sense. Ours is the Planet of the Insect. Insects inhabit almost every imaginable niche, tilling more soil than earthworms, living between the upper and lower surfaces of leaves and under our skin, inhabiting other insects, and boring out the wooden structures of our homes. They are the leading consumers of plants and the chief predators of plant eating insects and other insects. (Eisner and Wilson 1977, p. 3) They decompose humus and serve as food for vertebrates (animals with backbones).

Because of their small size, they opened up many more ecological niches than the vertebrates and are more diverse than all the rest of the animal kingdom combined. (fig. 3) There are an estimated 4 to 5 million species of insects with only about 1 million species classified and known to humanity. The greatest losses in biodiversity are occurring in the world of insects. There are untold numbers of insect species in rainforests inhabiting the innumerable niches they offer and millions of years of evolutionary complexity and disappearing as we continue to massacre our planet's rainforests. There are over a thousand species in a fair-sized backyard. The most successful form of multicellular life in terms of number of species are the beetles, followed by the butterflies and moths. (fig. 3) The proportion of fossil species to extant species is tiny in the insects compared to other animal species. We are truly living in the Age of the Insect.

Figure 3. On the next page is a pie chart showing the relative number of species in the major taxonomic groups of animals. Notice the dominance of four orders of insects; the beetles (Coleoptera), butterflies and moths (Lepidoptera), ants and related forms (Hymenoptera), and the flies and mosquitoes (Diptera). The proportional number of species remains about the same although estimates of the absolute numbers have changed since the chart was made in 1953. (From "The Relative Number of Living and Fossil Species of Animals," S. W. Muller and Alison Campbell, *Systematic Zoology* 3(4): 168-170 in Eisner and Wilson 1977 following page 4)

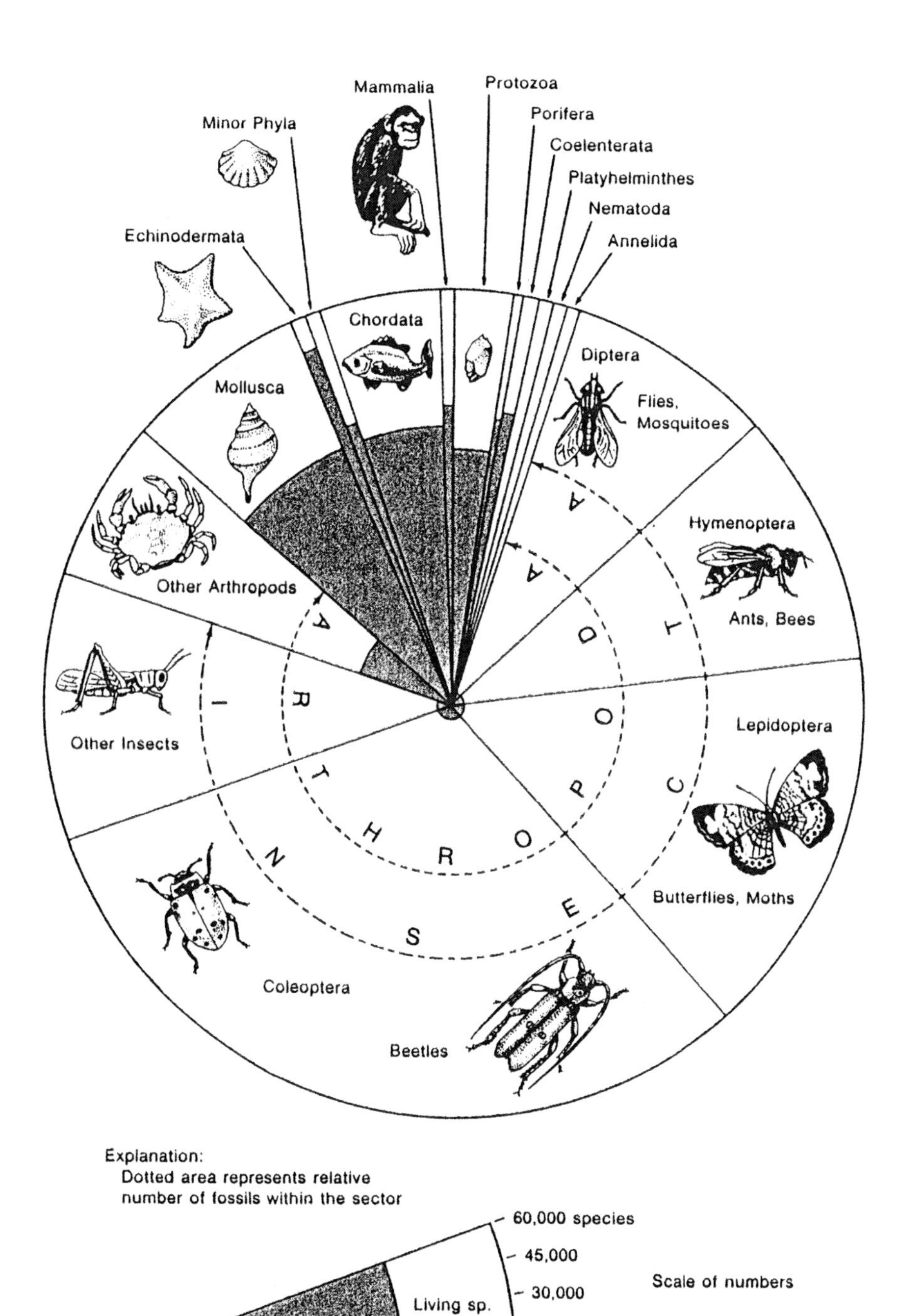
Mammalia
Protozoa
Porifera
Minor Phyla
Coelenterata
Platyhelminthes
Nematoda
Echinodermata
Annelida
Chordata
Diptera
Mollusca
Flies, Mosquitoes
Hymenoptera
Other Arthropods
Ants, Bees
ARTHROPODA
INSECTA
Lepidoptera
Other Insects
Butterflies, Moths
Coleoptera
Beetles
Explanation:
Dotted area represents relative number of fossils within the sector
60,000 species
45,000
30,000
15,000
1° = 3,000 species
Scale of numbers
Living sp.
Fossil sp.

Their large populations defy the imagination. It has been estimated there are 10^{18} insects alive at any moment. That's a billion billions. Every square kilometer of land surface has about 10 billion insects. Insects are exceptionally well adapted for life on land for more reasons than their small size, short life cycles, and tremendous fecundity.

1. They have a tough exoskeleton with a waxy coating that prevents water loss—a crucial factor for living on land. This rigid tubular casing protects internal parts and can be molded into an endless variety of shapes and retain an infinite variety of colors and patterns. This allows them to look like leaves, twigs and flowers and form such devices as stingers and ovipositors (egg-laying devices). The mouthparts can be molded for biting, piercing, cutting, lapping, sucking or forming instruments for building. Their amazing agility arises from an exoskeleton divided into segments joined by flexible membranes serving as hinges. (Eisner and Wilson 1977, p. 5, 6)
2. Because of their small size insects can breath by a system of air tubes (trachea) that penetrate from holes in the exoskeleton (spiracles) deep into tissues. This direct system allows rapid, efficient air movement which can service individual cells. Flight muscles in certain insects can generate more energy than any tissue in the animal kingdom, partly due to the trachea. (Eisner and Wilson 1977, p. 6)
3. Most insects undergo a complete metamorphosis which allows habitation of two entirely different ecological niches. The larvae (grubs, maggots and caterpillars) are eating and growing machines; the adults, usually winged, reproduce and disperse eggs. Wings are also useful for escape. Having the skeleton on the outside causes certain problems: to grow and increase in size, the smaller old skeleton must be shed. The transformation from larva to adult is made during the quiescent pupal stage when larval tissues are dissolved and adult organs are constituted from imaginal buds floating in the mush within the pupal case. There is no feeding or elimination of wastes in the pupal stage. (Eisner and Wilson 1977, p. 6, 7)
4. Insects display an incredible array of defensive mechanisms. They bite, sting, or eject noxious fluids. Most species have defensive glands of some sort. These are formed by invaginations of the body wall and therefore lined with impervious cuticular

membranes that provide appropriate insulation for the potent chemicals stored in the glands. Ants commonly eject a fluid composed of 20% formic acid, a powerful irritant. An Australian beetle larva discharges a fluid that emits hydrogen cyanide and some cockroaches spray a form of tear gas. Certain Carabidae beetles are "bombardiers," discharging a spray containing quinones they can accurately aim in any direction. The quinones are generated in an explosive oxidation reaction that liberates heat to create an ejected spray of 100° C (212° F). (Eisner and Wilson 1977, p. 10)

5. Males in the winged orders have an aedeagus, a skeletonized penis, to introduce sperm into the female. This allows sperm transfer in dry environments, sometimes in seconds, even in flight. The aedeagi of certain groups, including wasps, moths and flies, are incredibly complex, furnished with a strange array of spines, teeth, lobes, hooks and hairs. The variety allows a lock-and-key fit, serving as an isolating mechanism. This can be an important factor in the formation of insect species (distinct mating groups). Some species use the aedeagus to stimulate the females in courtship, others to hold unto the female until insemination is complete. (Eisner and Wilson 1977, p. 10, 11)
6. Insects have an amazing range of physiologies and behaviors. Many can hibernate or go into diapause (a suspended state of inactivity) if conditions are unfavorable as with extreme summer heat or winter cold. Some prevent themselves from freezing by producing ethylene glycol, a fluid put in car radiators. They have remarkably complex instinctual behavior for an organism with a tiny brain and nervous system. The tarantula wasp, for example, hunts down tarantulas that could kill it. The wasp has to sting the tarantula in a precise spot on the under side of its abdomen between two abdominal segments. Behind that spot is a ganglion that can be paralyzed by the wasp venom. The female drags the paralyzed tarantula to a pre-dug cavity in the earth and lays eggs in the tarantula. The wasp maggots develop within the paralyzed but living tarantula that the mother has buried.
7. Insects can be architects (bees, termites), potters (wasps) and paper makers (hornets). They make air conditioned colonies (termites), nets (caddisfly larvae) and shelters of clay, stone,

logs or galls. Some use jet propulsion (dragonfly nymphs) or cold light (fireflies) and chemical warfare. They solve aerodynamic and celestial navigation problems and employ elaborate communication systems involving chemicals (sex, alarm, trail following, and other pheromones), sound (cicada, grasshoppers and others), behavior (bee language) and light (fireflies).

8. Some of the most diverse and impressive behaviors are found among the social insects. Only 2% of the insect species are social insects but the power of large numbers working together performing specialized tasks makes these species incredibly successful. The social insects compose over half the biomass on the planet, while total insect biomass is 80% of the animal kingdom biomass. Social insects include the stingless bees, the social wasps, bees, ants and termites. Ants form 70% of the social insect biomass. In tropical areas they have four times the biomass of all the vertebrates.

One of the most interesting social insect species is a fungus cultivation ant that grows fungi on newly cut leaves carried deep into the soil. Their complex caste system is comprised of six different levels from cutters to gardeners. Specific activities include cutting leaves into small sizes, chewing the leaves to a pulp, mixing the pulp with enzymes, moving the fungal hyphae about, weeding the hyphae, etc. They excavate up to 40,000 pounds of soil to build a nest, becoming major pests that can strip a citrus tree in one night. The colony builds up to two million workers over several years living up to 20 feet underground. The queen lives 12-15 years and produces up to 150 million daughters who each live 1-2 years. The big guard ants are 200 times larger than the smallest ants and they can ward off vertebrates. A variety of pheromones (external hormones) are used to communicate and to control caste differentiation. (Klausnitzer 1987, p. 169, 170)

Insects feed on an immense array of foods in a variety of ways. Practically every species of land and fresh water plant is fed upon by insects. All parts of the plants are eaten, including roots, branches, fruit and seeds. Insects cause major destruction of our food crops before and after harvest. An estimated 5-10% of the world's food production is damaged or lost to insects annually, enough to feed 130 million people. (Elzinga 2000, p. 312) They eat our clothes and termites destroy our dwellings.

The pollinating services of insects are worth $3 billion annually in the United States and they provide about $4.5 billion of pest control by eating or parasitizing harmful insects. (Cornell University 2006, April 1) More than 425 plant viruses are transmitted by insects and more than 50 livestock diseases. (Elzinga 2000, p. 311, 304) It is estimated that 1 in 6 humans suffer from insect carried diseases. (Southwood 1997 quoted in Elzinga 2000, p. 297) Flies spread typhoid fever, cholera and dysentery; mosquitoes are vectors for malaria, yellow fever, West Nile disease and encephalitis; lice spread typhus and ticks carry spotted fever and Lyme disease.

III. The Imaginal, Mythical, and Symbolic Dimensions of Insects

How do these facts about insects impact the human psyche? Metaphorically, insects are much like the unconscious; virtually omnipresent and seemingly indestructible, displaying an immense range of habitats and forms. Incredible abundance, fecundity and sheer amount are characteristics of the unconscious as well as insects. Most insects are small and out of sight, working silently and unknown to humans—unless we develop an eye for seeing them and search them out patiently—again, much like the unconscious. Insects display incredible diversity and adaptability, reminding us of the immense creative energy of nature in which the human psyche partakes.

Insects are like concretized fantasies, images given three dimensional form in space/time. This realm of fantasy reminds us of Hermes, god of the unconscious and the imaginal, the psychopomp who leads into the underworld of the unconscious. The late Swiss biologist Adolph Portman professed that life's central purpose is display, a self-representation of a mystery. Portman suggested that an organism is not here just to preserve itself for a longer or shorter time, rather the preservation of the individual and the species serves the greater purpose of manifesting a being whose source is hidden from us. Self-preservation, species preservation and form-in-service-of-function are secondary, not primary factors of interest and consideration. (Portman 1982, p. 34, 35) (n 2)

Insect domains are often limited to micro-environments. The tree hole mosquito's larval world is in rainwater collected in a tree hole. The debris washed off the tree and the decaying wood in the hole provide nutrients for mosquito larvae to feed on. The bacteria and fungi that grow in this milieu are also larval food. The larvae may be parasitized by other insects which themselves may be parasitized (hyperparasites). An even smaller complex world with several interactions between species occurs within leaf galls. The galls are created by the plant's response to chemical irritants produced by the invading insect. To imagine being an insect in such environments is to be attuned to living in intimate relationship with other organisms and elemental forces operating over small areas. Ph.D. theses have been written on tree holes and leaf galls! Only the good scientist's, like a good artist's, inscrutable attention to detail and relationships can unravel the secrets of these little worlds.

The immense evolutionary age of insects and their biomass, diversity and domination of the land tell us they represent something fundamental about life. They are as old as plants and evolved together in symbiotic relationship with flowering plants (angiosperms). Many plant-insect associations have to be viewed as a single entity: the plant couldn't reproduce without the insect, the insect couldn't live without the plant. Insects are intimately plugged into the world, responding to subtle changes in moisture and displaying light tropisms (automatic responses or turning reactions to external stimuli like light). They are extremely sensitive to smells and chemical reactions, heat and cold, pressure and wind, and polarized light. Some even migrate by orienting to the earth's magnetic field.

Insects are at one with the basic cosmic factors, hard wired by instinctual responses to it. They are so sensitive to temperature changes because their small size gives them a large surface-to-volume ratio and they are unable to keep a constant internal temperature like humans. Hillman asks in his "Going Bugs" article, "Have bugs a cosmic knowledge, indicative of their accord with the order of things?" This cosmic connection is suggested in the dream of a 42-year-old woman as reported by Hillman:

> A huge praying mantis, maybe twenty feet high, says to me, "Are you a citizen?" I wake up screaming, "No!" (Hillman 1988, p. 62)

Hillman writes: "Has mantis (king of all creatures, according to the African Bushmen) come to her dream requiring a political awareness...Is she being shocked into cosmic citizenry, asked to become a citizen of an ecological civilization in which bugs are not put daily into holocaust?" (p. 62, 63) (see Appendix C for an example of a praying mantis as a woman's spirit animal)

The Navaho believe insects are at the primordial beginnings. They are the lowest level of things, not in the sense of being inferior, but in the sense of being the foundation of all things. The Navaho *"be'gotcidi"* insect figure is a Trickster: "the son of the Sun, who had intercourse with everything in the world" (Reichard 1950, p. 387 ff. quoted in Hillman 1988, p. 43):

> *Be'gotcidi* means "one-who-grabs-breasts," and details about him are too "dirty" to tell the anthropologist. "He got his name because he would make himself invisible, then sneak upon young girls to touch their breasts as he shouted...He also annoyed men...just as a hunter was ready to shoot, he would sneak up, grab the man's testicles, and shout...similarly when a man and woman were engaged in intercourse." *Be'gotcidi* is a "blond or red-haired god with blue eyes, dressed like a woman. He was in charge of insects, called them at will, and even sometimes appeared as a worm or insect." Once, when he was caught, hornets swarmed from his mouth, June bugs from his ears, mud beetles from his nose. Hornets stung all the other Gods, and *Be'gotcidi* swallowed all the bugs back then and could change himself into any sort of bug. (p. 43)

Hillman comments:

> The tales of this Bug-Lord present clear insight into the seeming spontaneity of insects, their cheeky irreverence for human intentions, their lordly power over us...Think... of the crazed state when last you tried to swat a gnat at night or demonically pursued a cockroach around the sink. (p. 43, 44)

Insects represent a certain indestructible, irreducible nature of life as revealed by this dream of an anxious woman in her 40's:

> Right in front of me in the middle of the street a young mother and three little girls are crouching by a low fire. They are burning insects. They sizzle and crackle and soon they die in the heat. But in the middle is a very big butterfly cocoon, a really tough creature. One of the girls tries to burn this too, but she doesn't succeed. It keeps coming out of the fire, staying alive. (Hillman 1988, p. 62)

Hillman wonders; "If indestructible, does this mean eternal?" A common fantasy is that insects will be the only survivors on the planet after a nuclear holocaust or whatever. Do they have this sense of foreverness because they can survive "the corrosive, exterminating powders of our rage? They show the tough autonomy of the will to live which is also an impersonal faith in life, as if a tropism permeating the cosmos." (p. 62)

The Greek word "psyche" means both butterfly and the human soul. The metamorphosis from an ugly, ungainly caterpillar into a beautiful butterfly during the quiet rebuilding stage in a chrysalis resonates with the human psyche's transformative abilities. Christians use the butterfly as a symbol of Christ's resurrection. The Lakota associate cocoons with *Wakinyan,* the spirit of the West as it manifests in generative powers which are initiated after old forms have been destroyed by the storms and lightning also associated with the West.

A positive mythological image of insects is seen in Demeter, the Greek goddess of crops and harvests, who taught humankind the art of cultivation and agriculture. She is called the Lady of Bees. One can imagine why Demeter's attributes are associated with bees: busy-as-a-bee activity leads to an industrious harvest of nectar. Honey production is facilitated by division of labor among the bees and working together, plus their architectural genius in the construction of the honeycomb. Crop yields are increased by the symbiotic relationship between pollinator and plant.

Bees have long affiliations with sacred traditions. The alchemists taught one how to become a bee, how to extract divine elements from others "and prepare a food in their hearts for angels." Muhammad thought the bee "was the only creature ever spoken to directly by God." (Lauck 2002, p. 136) The bee was associated with the soul in the Egyptian Ra cult and in India the gods Krishna, Indra and Vishnu are frequently portrayed as bees. Bees were linked with annual fertility rights and seasonal cycles because they disappeared in winter and reappeared in spring. Hives found hidden in the hollows of ancient trees or in rock crevices led to associations with the secret genesis of life hidden in the womb of the Earth. Bees are called Birds of the Muses in Asia and are believed in many parts of the world "to impart eloquence to a child of their choosing." (p. 137)

Honey was considered to be sacred in all ancient mythologies. It was believed to change into a mystical beneficial substance after being eaten, not unlike the transmutation of bread and wine in the Catholic communion. Symbolically it is associated with wisdom and sweetness from the mouths of the wise and strong. (Lauck 2002, p. 138)

The archetypal energy represented by the bee played an important role in the film *Ulee's Gold*, a healing sequel to *Platoon. Platoon* portrayed the horrors of the Vietnam war which are still reverberating through the American psyche. Ulee, played by Peter Fonda, lost his spirit in that war whose lingering shadow left him unable to be fully present with his family. His wife died, a symbolic statement about his feminine side and the psychological condition of too many Vietnam vets. His son was imprisoned for theft, his runaway daughter-in-law was wasted on drugs, and he was despised by the promiscuous teen-age granddaughter he was left to raise alone.

Rarely has a film so extensively portrayed the archetype of work—bees' most regaled quality. Ulee works himself to exhaustion trying to bring in the rich harvest of famed Tupelo honey during the brief three-week blossoming period in the Tupelo swamp. All aspects of harvesting honey are depicted and many comments are made about the difficulties of apiculture in America. Ulee's hard work eventually draws in the whole family to help him, and the son will carry on the family's apicultural tradition when he gets out of prison. We are reminded that the bees that work together in a colony are really one big family—all members arise from a single queen bee.

Just as bees and flowers have a symbiotic relationship, Ulee and his bees are intimately connected: "I take care of them and they care of me," Ulee said. Bees are one of the archetypal images for the spirit in nature; spirit as intelligence, order or ordering force in nature. It is Ulee's work with the bees and like the bees that helps bring order into his life and his family.

The dark side of bees is subtly dealt with. On a particularly difficult day after Ulee had made a daring rescue of his daughter-in-law and brought her home, she was having an ugly time coming off drugs. Ulee allows her daughter, eight-year old Penny, to skip school for a day and go out with him to work the bees. A bear had destroyed one of the hives and the bees had swarmed to a nearby branch. Ulee smokes them off into a new hive without getting stung. In the turning point of the film, Penny shows her drawing of this scene to her mother, metaphorically presenting the work Ulee is doing with Penny's mother. Penny describes the drawing to her mother:

> Sometimes the bees get confused and run away. They don't really want to leave. But they are happy when someone helps them find their way home. You have to keep calm

> and don't panic when they sting because they don't mean anything by it.

The biological basis of lashing out like angry bees in a fearful and destructive manner is described in Richard Goleman's *Emotional Intelligence* (1995). A very primitive part of the brain, the almond-sized amygdala, makes an initial quick assessment of a potentially threatening situation to determine if one is in trouble. The flight-or-fight response originates here. The many branches of the amygdale in the brain quickly rouse the person into action unless the frontal lobe can suspend action long enough to more carefully assess the situation. It takes training and maturity to allow the frontal lobe enough time to function and prevent what Goleman calls an "emotional high-jacking."

The ant is another highly successful social insect which shares many of the bee's admirable traits. As industrious as the bee, it is not bee-line direct and normally does not fly, but does move huge amounts of material one tiny bit at a time via tortuous ant freeways. "Go to the ant, thou sluggard, consider her ways, and be wise," it says in Proverbs 6:6. (n 3)

An ant figures in Mexican and Central American mythologies in association with corn, a plant of immense agricultural and religious importance in these cultures. Betty Fussell writes in her delightful book, *The Story of Corn*:

> For more than a thousand years, from the rise of the empire of Teotihuacan in A.D. 250 to the fall of the Aztec empire in A.D. 1521, Quetzalcoatl ruled the cradle of civilization in the New World. He brought corn to man after he had transformed himself into a black ant to steal a grain from the red ants who lived in the heart of Food Mountain. (Fussell 1992, p. 294)

A hermetic theft by a god brought something of great value over from the other side. It provided the symbolic and substantive base for the rise of New World civilizations—civilizations built on corn.

Most of our associations with insects are negative: things "bug" us, our computers need to be "debugged"; "What's bugging you is the nature of my game," says Lucifer in the Rolling Stones '60's hit, "Sympathy for the Devil." Our Western tradition is prejudiced against varmints, going back to the Bible and presented in Goethe's *Faust* (Part Two, 2, 1), "where a chorus of insects greets Mephistopheles, singing":

> O welcome, most welcome

> Old fellow from hell
> We're hovering and humming
> And know thee quite well.
> We singly in quiet
> Were planted by thee
> In thousands, O Father
> We dance here with glee.

Mephistopheles says, "This young creation warms my heart indeed." (Hillman 1988, p. 42)

Beelzebub, Lord of the Flies, loves insects "[who], like demons of the air and the night, and of hiding places in the earth, are his children. To consider the insect, to entertain its voices, is to listen to the devil," Hillman writes. (Hillman 1988, p. 42) The Judeo-Christian tradition used the connotation of Baalzebub as "Lord of the flies" to put down earlier religions. "*Baal*" meant "Lord" and "*zebub*" referred to a fly's humming or buzzing. (Lauck 2002, p. 45) "My Lord Who Hums" was once the "revered Philistine god Baalzebub—healer, 'Conductor of Souls,' and oracular deity." The Judeo-Christian identification of Baalzebub with putrefaction and destruction—"God of the Dunghill"—turned the important destructuring process served by insects into something despised and associated with sin, evil and pestilence. (p. 46)

Death-and-rebirth transformations associated with insects are archetypically related to the goddess cults that the Judeo-Christian tradition suppressed. (see Appendices G and H in volume 3 of *The Dairy Farmer's Guide*) The cults honored the rhythmic nature of the seasons as a model for psycho-spiritual transformation by descent, death and rebirth. (Lauck 2002, p. 61) Acceptance and surrender rather than heroically battling these forces allows one to shed old identities and be reborn from a deeper source. Joseph Campbell pointed out, "it is always the disgusting and rejected creature that calls us to some undertaking, some spiritual passage." (p. 81) As Joanne Elizabeth Lauck describes it in her beautiful book on insects, *The Voice of the Infinite in the Small*:

> For without openness to this side of life, to the decay-sensing, death-wielding forces of transformation—the flies' domain—we will be worked on invasively, pitilessly, against our personal will, because our will, what we want, is too small. (p. 61)

The Navajo believe Big Fly is the voice of the Holy Spirit who mediates between them and their gods. This revered mentor, often appearing in sandplay paintings, counsels and prognosticates while sitting on or behind a person's ear. (Lauck 2002, p. 55) For the Hopi, the well-known flute-playing Kokapelli is a personification of the local assassin or Robber fly. (p. 55, 56)

The roots of the Western negative view of insects in dreams appear in the first book on dream interpretation by Artemidorus ca. 150 AD:

> Whenever ants crawl around the body of the dreamer, it portends death, because they are cold, black and the children of the earth. [lll: 6] Bugs are symbols of cares and anxieties...discontent and dissatisfaction. Gnats...signify that the dreamer will come into contact with evil men. [lll: 8]...if there are many lice...it is unpropitious and signifies a lingering illness, captivity, or great poverty...if a person should awaken while he is dreaming that he has lice, it means that he will never be saved. [lll: 7] (Artemidorus 1975 quoted in Hillman 1988, p. 42)

Nero dreamt he was covered with winged ants after he had murdered his mother. Saintly Celia was crucified on an anthill in T. S. Eliot's *The Cocktail Party*. Freud viewed bugs in dreams as vermin representing unwanted little brothers and sisters who plague us. The spider is seen as the negative mother or negative self image in Kafka's *Metamorphosis,* Jean-Paul Sartre's *Les Mouches*, and by Jung. (Hillman 1988, p. 42)

Jung thought that insect dreams might mean symptoms could be lodged in the vegetative (autonomic) nervous system that unconsciously innervates basic body processes like digestion, heartbeat, etc. Like insects, the vegetative nervous system is not susceptible to understanding or willing. The vegetative level of the unconscious is associated with Hermes' realm as discussed in volume 3 of *The Dairy Farmer's Guide.*

Although insects often appear negatively in dreams, the intent may be to raise consciousness and transform the dreamer. Bug-in-the-wound dream motifs are symptoms which attempt to cure the condition that required the symptom. Since most insects serve a destructuring and recycling role in nature, they are naturally attracted to what is malfunctioning. Insects in dreams can show us what is not functioning in our physical or psychic health, alerting and sensitizing us to what needs attention, transformation or elimination. (Lauck 2002 p. 250)

Insects often exhibit intentionality in dreams as illustrated by three dreams presented by Hillman. (Hillman 1988, p. 45) A young woman living at home with an over managing mother began analysis with this dream:

> I have on a pretty dress and moths and insects keep flying to it like a magnet. They keep flying onto me and eating the dress.

A depressed, immobilized man of thirty who was occasionally a barfly dreamt:

> Sitting at a bar drinking. Great big insects appeared and started jumping up at me.

He later dreamt:

> I was walking down the street somewhere and saw on the pavement a swarm of black spiders which scattered away from me as I walked along. I stopped to look and as soon as I stood still, they mounted force and came toward me in one spearhead formation.

The intention of the insects in these dreams can be seen from their effect on the dreamer. Hillman comments:

> First we see the movement of creatures: "flying onto me," "jumping up at me," "toward me in one spearhead formation." They intend the dreamer...At least they want the dreamer by closing in on her/him...*What* they specifically want can be determined only by the image in which they appear. For the young woman, it is her dress (a red one, chosen by her mother to attract young men; but she was shy, slightly anorexic, and red did not suit her feelings). For the man at the bar, the jumping huge bugs interfere with his sitting and drinking. (Hillman 1988, p. 45)

The dreamer often realizes something in the dream: insects often arouse, activate, alert us—traits associated with consciousness. Jung spoke of consciousness in the unconscious as a "light in nature." (CW 8, ¶ 388 ff) Insect dreams can disturb us into awareness and the disturbance often comes suddenly, announcing a crisis point in the dream. It becomes the catalyst for invading and breaking up a pattern, allowing something to start depending on how the dreamer reacts in the dream. (Hillman 1988, p. 47) The disruptive nature of dream insects has overtones of Jung's concept of God:

> To this day God is the name by which I designate all things which cross my willful path violently and recklessly, all things which upset my subjective view, plans and intentions and change the course of my life for better or worse. (Jung 1961b)

Our response to insects in dreams is not unlike our response to insects in waking life: a panic reaction leading to an overkill desire to eradicate. (n 4) The Dalai Lama addressed this response when asked what he thought was the most important thing to teach children: "Teach them to love insects," he said. This will help cultivate tolerance and respect for differences within ourselves and with others. (Lauck 2002, p. 114)

Hillman lists four frightening fantasies attributed to insects which may be behind the over-kill response. The first fear is that the incredible numbers of insects "indicate insignificance and worthlessness as individuals." (Hillman 1988, p. 59) Our sense of being a unique and unified individual is threatened. Insect dreams are usually

> interpreted as signs of fragmentation and the lowering of individualized consciousness to an undifferentiated, merely numerical or statistical level. Then the invasion of insects in a dream indicates psychotic dissociation and the loss of centralized control. Eradication, then, is an "anti-psychotic." (p. 59)

The source of the psychosis may really be in the defensive unity of the eradicator.

The issue, Hillman explains, is how we regard multiplicity: "Once we imagine multiplicity through the single lens of a unitary human being, and conceive wholeness as oneness, the insects become the active embodiments of the Many Against the One." (Hillman 1988, p. 59) In contrast, the swarm shows unity and multiplicity at once. Hillman notes:

> The anthill is also a community, the active embodiment of...(social feeling), and the crowd of insects demonstrates wholeness...as a busy, buzzing body of life, going every-which-way at once. The swarm redefines wholeness as cooperative complexity. (p. 59)

The anthill provides an ecological model of the psyche where the ego we associate with in dreams is just one of the many people we live with within. Jung called these dream people the "little people" within, (CW

8, ¶ 209) and Richard Schwartz, developer of Internal Family Systems, models the psyche as the ego living with a tribe or community of people within. The monotheism of the Judeo-Christian tradition provides an archetypal base for a dominating, central ego nature in contrast to the more inclusive archetypal base offered by polytheism. As developed in volume 1, Jung described Christianity as a dualistic religion—God and the Devil. A domineering force tends to demonize what it tries to dominate, then attempts to eradicate "evil."

Ant colonies live near "the edge of chaos," a important concept in complexity and chaos theories. Individual ants behave in a chaotic manner, but when enough ants interact or communicate with each other they shift into a "super organism" state, a rhythmic, orderly state. Entering this state depends on a certain density of ants per inhabited area and they seem to be able to regulate their numbers to achieve that density. This puts them between order and stability on one hand and chaos on the other. (Goodwin 1994, p. 287 referenced in Lauck 2002, p. 106) Mathematical models describe this as the optimal state for open complex systems, the state where new forms and behavior can emerge and evolve in response to an environment. (see volume 3, Appendix A) A little chaos is good for loosening up old forms to allow for transformation and reorganization to occur. This is a model for the ego in relation to the unconscious, the individual in relation to the collective, and groups in relation to each other.

A second frightening insect fantasy revolves around the monstrosities and the myriad of bizarre forms insects can assume. An insect is "without the warm blood of feeling" and we use terms like "bug-eyed, spidery, worm, roach, bloodsucker" to contemptuously characterize inhuman traits in people. (Hillman 1988, p. 59) Hillman writes:

> Insects in dreams suggest the psyche's capacity to generate extraordinary forms almost beyond imagining...These inhuman monstrosities show the reactive potential of the psyche beyond its humanistic definitions. They bug us out of ego psychology, out of humanisms. (p. 59) (see volume 3, Appendix K)

A third factor is the autonomy of insects: we can crush, burn and poison them but they wouldn't submit. They have other intentions, competing with us for home, food, and living space. They bug us with "the autonomous nervous systems' persistent symptoms" and an autonomy that eats wounds into me, stings me into rage, reveals my rot and holes, and drives me crazy. "The 'me' believing itself in possession of autonomous free will is relentlessly pursued by the...(unconsciousness)" on which it rests and nests in. (Hillman 1988, p. 60) Jung wrote about a theologian who was frightened by a dream of the wind stirring the surface of a lake in a dense woods. The wind represented an unseen presence that "bloweth where it listeth...a numen that lives its own life." (Jung 1961a, p.141) That the unconscious can have a will of its own and act autonomously is frightening to a strongly centralized ego believing that the conscious world is the only safe human domain.

The fourth fear is of parasites; a fear of being eaten up from within, of having our personalities altered by an alien power that we call our complexes, our hang-ups. (Hillman 1988, p. 60)

Insects in dreams can represent the return of the repressed. They are often associated with dirt, beings from under the earth, toilet bowls, manure piles, pubic areas and rotten meat. Hair and the lower part of the body are often affected. Spiders and centipedes have been interpreted as anal symbolism, repeating "the idea of the bug as the evil outcast, smelly, sulfuric, of the Devil." (Hillman 1988, p. 68) As Hillman describes it:

> The low evaluation corresponds with the bug's underground surreptitious concealment. Hidden, buried, interior, appearing at night through small openings in day world structures—these attributes suggest the underworld...Maybe they refer neither to the morally repressed (evil), the esthetically repressed (ugly), nor the primordially repressed (death), but to the chthonic Gods, especially Hades, who emerges through—and whose intentions live in—those holes we feel as wounds. (p. 68)

The bug bite then is an underworld wound, indicating the presence of death "and the cosmic urge of desirous life to live." (p. 68) Dostoevski's Karamazov brothers were described as insects, and it is "the insects to whom God gave 'sensual lust'...[that] will stir a tempest in your blood." (F. M. Dostoevski, *The Brothers Karamazov,* quoted in Hillman 1988, p. 40)

The alchemists praised the stone that the builders rejected, echoing wisdom tradition teachings that the source of renewal comes out of the humble and the small. Lauck, in her book on the insect-human connection, called for individual and cultural recognition of insects as "the gift hidden in the midst of the Shadow." (Lauck 2002, p. 273) This ecopsychological perspective accepts the dark and difficult aspects in our psyches and in nature, not demonizing them but recognizing their paradoxical role in growth and transformation.

Hillman tells us that our dreams of insects reveal they have something to teach us about basic elements of the natural world and the Gods in diseases:

> They demonstrate the intentions of the natural mind, the undeviating faith of desire, and the urge to survive...
> To survive as they survive, we must utterly transform the shapes of our thought, as they risk all in their transformations...
>
> We keep the Gods alive with flesh, our animal flesh...The bugs in our dreams pierce into us, bite and draw blood, reminding us we too are meat...The Gods become diseases; ourselves infested by Gods, forced to religion by bodily sensations...
>
> [In dreams], forgotten pagan polytheism breeds in animal forms. In those animals are the ancient Gods...still there... [as] ikons of our dreams and in the vital obsessions of our complexes and symptoms, the little bugs indestructible. (Hillman 1988, p. 70, 71)

IV. Synchronicity and Insects

I, as an entomologist and Jungian analyst, have been particularly interested in the many synchronicities involving insects. This is Hermes' domain, given his association with synchronicity, the vegetative nervous system, and the bee oracle received from Zeus and Apollo (volume 3). An example of a woman's synchronistic experiences with the praying mantis is given in Appendix C. My personal best experience happened while training at the Jung Institute in Zurich. In a dream I noticed there was a gaping wound down to the bone at the base of my right index finger. No blood was present but there was a white insect larva about a quarter of an inch long on the wound. Upon awakening from the dream, I remembered that a friend was doing a thesis on the archetype of the wounded finger, developing the idea that the wounded finger represented a too conscious approach to the creative process. (The Greek name for finger is "dactyl" and the dactyloi helped Mother Earth give birth to Zeus, a creative process) As often happens at Jung

Institutes, people collect dreams associated with their thesis topics, so I promptly passed the dream on to my friend. A few months later I went by train to Zurich with my young daughter to buy a Christmas tree. We found an empty seat on the train and it happened to be across from my friend. I pulled off my stocking cap, my right glove, and then my left glove. I was astonished to see at the base of my left index finger a green insect larva a quarter of an inch long. It had barely registered that as I was pulling off my right glove there was something at the base of my right index finger. My friend and I immediately recognized the synchronicity. I believe that the larva had been at the base of my right index finger but I didn't notice it because I was preoccupied with greeting my friend and settling in. My second chance to recognize the synchronicity came when the larva somehow ended up on my left index finger. As an entomologist I fully appreciated the improbability of finding a live insect larva in Switzerland in December, let alone finding it inside my glove and in such a meaningful place for my friend and me. This meets my strictest definition of a synchronistic event: a meaningful coincidence improbable to the point of being miraculous. Synchronistic events usually occur around archetypal constellations, in this case the archetype of the wounded finger which I shared with my friend. He was doing his thesis on the wounded finger because he himself had a powerful dream about a wounded finger. We had both entered the Jung Institute with strong scientific backgrounds.

An entomologist's familiarity with insect shapes and life styles allows them to recognize synchronistic events that a non-entomologist might miss. Two such examples occurred at the same sacred Native American site on the grounds of Mendota State Hospital beside Lake Mendota in Madison, Wisconsin. The southern part of Wisconsin and bits of the adjoining states include the entire domain of a culture that over a thousand years ago made earthen mounds in the shape of birds and animals. French fur trappers, the first white people to come through this region, called them "effigy" (image) mounds. The mounds are arranged so as to tell stories on the ground and usually contain a few mounds aligned to the solstices, equinoxes and/or the 18.6-year moon cycle. (Maier 2001; Merritt 2008) Important tribal members were often buried in the mounds with directional, solstice or equinox alignments, probably to launch the soul on a spiritual journey after death back to the heavens. (Maier 2001)

The first synchronicity at a mound site occurred when I was in a group of people listening to Dr. Maier. He was describing how a tree that had grown into the side of one of the mounds he was standing beside had been blown over in a big storm. Its upended roots removed the end of the mound, exposing skeletons and grave artifacts within. Burial mounds are protected, so this natural event is the only way something hidden in the earth could have been brought to light. As my friend was describing this, I noticed that the morning sun was glistening off something at the base of a nearby tree. When I recognized what insect it was, I knew a synchronicity had occurred.

What insect would be related to the story being told at that moment? It was the cast off exoskeleton of the last nymphal state of a cicada, whose cousin in known as the 17-year locust. This family of insects can spend 17 years or more underground in an immature nymphal state, sucking sap from the roots of trees. They emerge en-mass and attach themselves to the bases of trees and other objects where their exoskeleton splits down the back to allow the emergence of a winged adult. The males make a loud buzzing sound to attract a female beginning about mid-July in Wisconsin. The archetype of resurrection emerged: something buried in the earth for a long time had come to light, mirroring Dr. Maier's explication of the long-lost meaning of the mound burial. The adult cicada's emergence from its immature form that lived underground would be equivalent to a butterfly emerging from a chrysalis, a Christian symbol of the resurrection mentioned before.

Another synchronicity occurred a few years later at the other end of this large sacred site. Dr. Maier was showing my friends an eagle effigy mound near Lake Mendota with a 624-foot wing span. While standing at the end of this mound I asked Gary if it was true that Mendota means "the lake where the Indian lies." As he started to answer, I noticed there was an insect on his left cheek that I thought at first was a mosquito, but it was too big. Gary replied that Mendota means "the lake where the Indian mates and dies." I looked closer at the insect and recognized a synchronistic event. An entomologist would immediately think of one insect associated with water that mates and dies. Mayflies emerge en-mass from the river-bottom environment of their youth and cling to objects while splitting open their exoskeletons. This allows a flying adult to emerge. Adults live for only a day to mate, lay eggs, and die. Indeed, the insect on my friend's cheek was a mayfly.

Synchronistic events often occur around death, one of the most powerful archetypes, and I had a synchronistic experience around the death of my mother that included an insect. It occurred at the Sundance on the Rosebud Reservation in South Dakota in June following my mother's death in January. I wanted to deal with my mother's death in the deepest possible way and the Sundance provides a safe and sacred container for grieving. I pay particular attention to insects and their behaviors when I'm at the Sundance because it is such an archetypically charged ceremony. While observing from my usual position in the northwest section of the arbor, I noticed a Tiger Swallowtail—those big, beautiful yellow butterflies with the long "tails"—come flying towards me from the direction of the Sundance tree in the center. Many cultures believe butterflies and moths are the souls of the departed and bring messages from the dead. The butterfly flitted off to my right, the west, and disappeared. Soon it was flying past me again, going from my right to my left, then flew out through what is called the north gate in the Sundance circle. (A gate consists of two flags spaced about four feet apart) I recognized the significance of the butterflies' flight. Lakota associate the west with death and believe that when a person dies their spirit moves to the north to the spirit trail, the starting point being the southern end of the Milky Way guarded by an old woman. If they had not lived a good spiritual life they were not allowed to enter the trail. The stars in the Milky Way are believed to be the campfires of those moving north to their final destination. One more thing: my mother loved butterflies. As a remembrance I have on my alter a yellow butterfly broach she used to wear.

The Sundance experience encapsulated a sense of self woven together from several strands in my life. I particularly associate my mother with me growing up on a dairy farm. She was raised on the same farm and was a salt-of-the-earth type woman. I became a doctor of philosophy of the world of insects I discovered on the farm. "Psyche" as the Greek word for "butterfly" *and* "soul" combines my entomology and Jungian identities. The Sundance has been a place of spiritual renewal and healing for me within an indigenous cultural and natural context. Synchronicity experiences have moved me, and everyone who has honored them, into a profound and mysterious world beyond science and into a cosmos where "we are all related" at many levels—ecopsychology in its broadest and deepest dimensions.

NOTES

Chapter 1: An Archetypal View of the Midwest Environment

1. Most of the Midwest is part of the Mississippi River drainage basin except for a portion of the northeastern fringe of states. A subcontinental divide helps create the Great Lakes drainage basin that drains east through the St. Lawrence River into the North Atlantic.

   Within the Great Lakes basin and forming part of the subcontinental divide is the Niagara Escarpment, the remnants of a 400 million-year-old tropical barrier reef formed during the Silurian Age. The reef delineated the Michigan basin of an inland sea in a circle through Lake Huron, upper Lake Michigan, eastern Wisconsin and northern Illinois, Indiana, and Ohio, with the eastern end being the ledge of the Niagara Falls. The ledges in Door County and along the western side of Lake Winnebago in Wisconsin are prominent features of the escarpment. Our dairy farm was over a submerged portion of that reef and I once found a piece of petrified coral while picking stones before corn planting.

2. Corn has more varieties (over 12,000) than any other crop because it hybridizes so readily. (Fussell 1992, p. 93) The most successful corn hybrids are made by "double crossing" corn. "Hybrid vigor," first mentioned by Darwin, is achieved in corn by cross breeding inbred strain A with inbred strain B to produce AB and another combination of strain C with strain D to produce CD. The two combinations are then crossed (to get a "double cross") to produce an ABCD hydrid. Cross breeding is done by covering the developing ears with paper bags and snipping off the tassels which are then used to hand-fertilize the ears. Every summer a small army of youths earn spending money by toiling for a couple weeks in the hot, humid weather of corn breeding season. (p. 71)

3. The Corn Palace in Mitchell, South Dakota is a survivor of the once popular practice of city and state boosters of making corn palaces at National and International exhibitions. The Mitchell Corn Palace has huge outside panels annually decorated with various scenes using a palate of a hundred thousand corncobs and two thousand bushels of various grasses and weeds. Varieties of corn are specially grown for particular colors. (Fussell 1992, p. 317)

4. While indigenous corn cultures in the Americas had hundreds of corn dishes (Fussell 1992, p. 174), we have a restricted diet of popcorn (thank you Orville), sweet corn, creamed corn, corn flakes, polenta and corn

dogs; corn nuts, corn bread and corn mush; hominy and grits, pozole and taco shells, tortillas and tamales. (Fussell 1992, p. 167-248)

5. Edward Janus' book, *Creating Dairyland* (2011), illustrates the *I Ching's* statement, "Care of the cow brings good fortune," (Wilhelm 1967, p. 119) by chronicling the development of the dairying industry in Wisconsin. The early history of the state was marred by land speculation and wheat farming—a crop that mined the soil. After the Civil War, men like W. D. Hoard became strong advocates for developing a dairying industry, believing it would bring prosperity to the farmers and the state and improve the character of humankind:

> Hoard and the other dairy founders would propound a program for the improvement of farmers that was the embodiment of the emerging progressive spirit of the time, a spirit born of faith in human intelligence and the social order rationality would create. From this ideology grew a practical program for the wide-ranging improvement of farming practices, attitudes, and education—the beginning of the Wisconsin Idea of Dairying. (Edward Janus, 2011, *Creating Dairyland,* Wisconsin Historical Society Press: Wisconsin, p. 17)

The geography and people of the state were transformed by the intense nature of dairy farming, rebuilding the soil with cow manure, the tremendous increase in productivity, and the use of soil conservation techniques resulting from discoveries in university-sponsored research. The modern agribusiness model slowly evolved through the education of farmers in university-sponsored Farm Short Course and county extension agents, improvement in cattle genetics and feeds, development of dairy products and sanitation standards, and the cultivation of new markets throughout the world. The motto of the University of Wisconsin, "The boundaries of the University are the boundaries of the State," has it genesis in the university's involvement in developing dairying in Wisconsin. In the end it is the cow herself who had the greatest transformative effect:

> Cows require kindness, routine and predictability, and long-term investments for their well-being and our profit. Dairy farming must be dedicated to caring well for other living beings and for the resources that feed them. This devotion is the basis of a dairy farmer's business and of our conservation ethic, and over the years thousands of young men and women have carried this ethic off the farm and into the professions, cities, and politics of our state. This ethic of caring is a pervasive influence in our lives. Cows have shaped us. (Janus 2011, p. 3)

6. One of Jung's ideal images of the archetypal feminine was of the woman as a farmer:

> Our women are uprooted. They are no longer identical with the earth—perhaps identical with a flat but not with the earth. But one sees it under primitive circumstances. I have seen it in East Africa, though I don't know how long it will be before the missionaries succeed in destroying the original order of things. The woman there owns the *shamba*, the plantation. She is identical with her estate and has the dignity of the whole earth. She is the earth, has her own piece of ground, and so she makes sense. She is not up in the air, a sort of social appendix. (Jarrett 1988, p. 1282-1284)

7. "Minnesota: A State That Works," *Time* 102 (Aug. 13, 1973: 24-35), p. 35 quoted in Shortridge 1989, p. 144.

Chapter 2: Seasons of the Soul

1. Peter Redgrove, a weather sensitive, writes in *The Black Goddess and the Unseen Real*, "Some 80% of the sensitives can actually predict the weather." Women are more sensitive than men, probably due to their production of less stress-resistant hormones. (Redgrove 1987, p. 91)
2. Atmospheric electricity that precedes the rains by up to 150 miles is what most directly affects people and not excessive heat and humidity that precede a storm. The effects reach down to the glandular level by measurably affecting "male and female sex hormones, thyroid, growth hormones, insulin and calcium metabolism hormones." There is an over-production of the hormone serotonin, produced in the brain and the intestines, in response to atmospheric electricity. Redgrove notes, "[This] can result in insomnia, irritability, migraine, vomiting, palpitations, and a feeling of electric currents on the skin and hair and many other symptoms, as Coleridge knew." (Redgrove 1987, p. 90) Atmospheric electricity correlates with death rates and muscle reaction times, affecting sport and work activities. (p. 91)

   Weather alters atmospheric fields, producing low electrical gradients in clear air and high gradients in fog. "There is a low gradient above the sea and high positive and negative gradients fluctuating in snowfall and thunderstorms, which must account in part for the depression and fascination caused by these kinds of weather." (Redgrove 1987, p. 90) "The ionization of the ground alters as each cloud passes over, and [the] lungs breath this charged air." (p. 112)

   Solco W. Tromp, who did monumental studies and compilations of biometerology, spoke of a "wonderful web of electromagnetic forces which seem to regulate all living processes on earth." (S. Tromp, 1949, *Psychical Physics*, Elsevier: New York quoted in Redgrove 1987, p. 93)

3. Those near death may make miraculous recoveries and there may be a temporary drop in the death rate after storms. Students get higher marks and factory workers work faster just after storms. (Redgrove 1987, p. 82)

4. Data prove, for example, that seasonal weather changes affect interest in reading. (Redgrove 1987, p. 82)

   Coleridge acknowledged the essential relationship between his weather sensitivity and his poetic sensibility. After close contact with dampness in nature or pre-storm conditions, Coleridge suffered from depression. "When repressed, his dark or animal senses showed themselves as illness, and when the repression was overcome by working it through in poetry, illness was replaced by insight and the convalescent condition, which he regarded as a state of genius." (H. House, 1962, *Coleridge: The Clark Lectures, 1951-1952,* Rupert Hart-Davis: London, p. 38-39 referenced in Redgrove 1987, p. 84) Coleridge had an extrasensual skin, a female secondary sexual characteristic, with a consciousness accompanying it that he described as being infinitely mobile and reflective of everything. His skin reflected the weather, which stimulated his poetic imagination, and his stomach was in instant sympathy with his skin, feeling uncomfortable in damp and stormy weather. (Redgrove 1987, p. 85, 86) Coleridge sought a symbolic language to express this connection to nature that he, and everyone, experiences at a deep, physiological level. Nature provided Coleridge with "'the Shaping Spirit of Imagination'" by which he could discover the hidden connection of the body with nature and the weather. When his bodymind was released by the passing of a storm, his "imagination [rode] on the wind over the world, as though it were the wind's perceptions, human and natural at once, simultaneously embodying themselves in superb poetry." (p. 88)

5. Weather sensitives "are so violently affected by weather changes...that it can easily become a clinical problem, either psychiatric or organic." (Redgrove 1987, p. xi) Redgrove, quoting William A. R. Thomson, reminds us that "climatology has been the basis of the practice of medicine from time immemorial...The doctor had to be 'a meteorologist if he was going to do justice to his patients.'" (William A. R. Thomson, 1979, *A Change of Air: Climate and Health,* Charles Schribner and Sons: New York, p. xi-xii, quoted in Redgrove 1987, p. 80) Yet bioclimatology is scarcely taught in medical schools. During the Enlightenment weather was dismissed and repressed as a major concern in Western philosophy and literature. Arden Reed correlated the changing Enlightenment attitude towards weather with the development of modern science which depends on exact measurements in totally regulated environments, "and this necessarily excluded weather as a variable that could be controlled." (Arden Reed, 1983, *Romantic Weather,* University

Press of New England: Hanover and London, p. 4 quoted in Redgrove 1987, p. 82) Redgrove comments:

> It is difficult to forgive our forefathers for the intellectual dishonesty of ignoring what was not expedient to understand, while professing reason. Since weather was natural and could not be controlled, it was ignored...and millions suffer because part of the causative picture of their illnesses has been repressed. (Redgrove 1987, p. 83)

6. The medical community does the weather-sensitives a disservice when they treat for mental or physical problems "when they were in fact 'communing with nature,' and were not given images by our society to celebrate or express this communion." (Redgrove 1987, p. xiii)
7. When one consults the *I Ching* using the yarrow stalk method, one counts off by four sticks which represent the four aspects of any cycle of change.
8. Jungian analyst Patricia Berry describes a Demeter upper world as a "realm of concrete, daily life, devoid of spiritual values." Persephone, the Black Goddess and Lost Daughter, is the underworld daughter who carries "the sense of essence and the dark (and beneath the dark)" (Berry 1982, p. 15-20, 31 in Redgrove 1987, p. 155):

> Apprenticeship to the White Goddess of the full moon, ovulation, physical fertility, is served in the upper world, with Demeter. Jung calls this 'paying the debt to nature' before the work of individuation starts.
>
> In the underworld one is among the essences, the invisible aspects of the upperworld. (Redgrove 1987, p. 155)

To perceive concrete objects so deeply that one perceives them "as germinations of the realm of Hades" is to have a soul-full Demeter/Persephone perspective. "The concrete natural world...is [then] the very way and expression of soul." (Berry 1982, p. 15-20, 31 quoted in Redgrove 1987, p. 156) For more on the Black Goddess, see volume 3, appendix H.

9. Jack Santino defines ritual as the "repeated and recurrent symbolic enactments, customs and ceremonies that are often carried out with reference to the sacred, or at least some overarching institution or principle." (Santino 1994, p. 10)
10. Pan is the most difficult of Hermes' children to examine. He was "fantastic to look at, with goat-feet, and two horns, very noisy but laughing sweetly." (Boer 1970, p. 67) Exploration of Pan takes us into the darkest realms of the Western psyche. Raphael Lopez-Pedraza notes that "Pan was at the center of the repression of the pagan Gods...Pan and the

complexes around him became equated with evil and the imagery of the Devil." (Lopez-Pedraza 1977, p. 87)

Repressed energy causes the gods to "appear at the core of our complexes, sometimes autonomously in neuroses and psychoses, and in physical illnesses" and symptoms. (Lopez-Pedraza 1977, p. 80) Lopez-Pedraza recognizes the supreme challenge of Pan *and* his vital importance:

> Pan [is] the missing link to the physical body (p. 80)...[He] represents the furthest possibility for the psyche, a frontier where the personality can either find some natural self-regulation or go into permanent madness. (p. 87)

Pan is the god of panic, creating "the most panic when his image is presented under the historical disguise of the devil." (Lopez-Pedraza 1977, p. 82) He is god of the nightmare, which necessitates seeing that terrifying experience as an epiphany of a God. (p. 83) As the god of masturbation,

> Pan gives a frame of reference for the whole gamut of masturbatory fantasies...The repeated sexual image—reflecting the part of man's nature which does not change—can either be accepted, and so bring a deeper insight into the personality, or can continue obsessively with no insight. Moreover, masturbation is also the field where new sexual imagery first occurs...The part of man's nature which moves—enables the self-detection of new psychic movers...Pan, as have all the Gods, has the two-sidedness of sickness/healing. (p. 83)

Lopez-Pedraza emphasizes it is important for psychotherapists to realize that human nature has two parts: "a part that does not change and a part that moves...[To] detect what...does not change, [allows one] to localize our psychotherapeutic aims in that part which moves." (Lopez-Pedraza 1977, p. 89 note 18)

In classical Greece neither Pan nor any of the chthonic gods "carried a projection even remotely akin to the Christian one." There were devils and evildoers in a later Greek tradition: "these were Titans made of iron and steel." Lopez-Pedraza adds, "All the Gods and Goddesses each have their own style of destruction and killing." (Lopez-Pedraza 1977, p. 87)

11. For the seasonal dimensions of the church year see Christopher Hill, 2003, *Holidays and Holy Nights: Celebrating Twelve Seasonal Festivals of the Christian Year*, Quest Books: Wheaton, IL.

Chapter 3: Planet of the Insect

1.

> "We assume...," says Frankfort, "that the Egyptian interpreted the non-human as super-human, in particular when he saw it in animals—in their inarticulate wisdom, their certainty, their unhesitating achievement, and above all in their static reality. With animals the continual succession of generations brought no change...They would appear to share...the fundamental nature of creation": its repetitious, rhythmic stability. Each animal confirms that living forms continue; they are eternal forms walking around. An animal is eternity alive and displayed. Each giraffe and polar bear is both an individual here and now and the species itself, unchanging, always self-creating according to kind." (Frankfort 1961, p. 8-14, quoted in Hillman 1982, p. 321)

2. Swiss biologist Adolf Portmann maintained that

> animal life is *biologically aesthetic*: each species presents itself in designs, coats, tails...songs, dances....The aesthetic is rooted in biology...The coat of an animal is phylogenetically prior to the optical structures necessary for seeing the coat. To show is primarily to show—secondarily to be seen...There are symmetrical markings on primitive oceanic organisms that bear no useful purpose...Sheer appearance as its own purpose recalls the original aesthetic meanings of *kosmos*—ornament, decoration, embellishment, dress. An animalized cosmology restores the aesthetic to primary place. To restore this aesthetic sense of cosmos therefore requires giving first place to animals. (Hillman 1989, p. 27, 28)

3. For ant stories and a review of the mythological and symbolic dimensions of the ant see Lauch 2002, p. 91-111.
4. Pink Floyd's *Grandchester Meadows* can be interpreted as a song about the disastrous consequences of our extermination approach to insect pests and unconscious symptoms. (see Appendix D: Pink Floyd and the Fly in Life's Ointment)

APPENDIX A

Sacred Landscape at Strawberry Island

Herman Bender (2009) describes how a physical landscape can mirror a spiritual landscape, establishing both a center and sense of place for a tribe and creating a path to paradise during certain precise alignments of heaven and earth. Bender's work with the Lac du Flambeau band of Chippewa in northern Wisconsin illustrates how a tribe transcribed its symbolic maps onto a new environment when they moved westward into Wisconsin. Together with establishing sacred lodges, the transcription was done as "a means of both consecrating and claiming territory as their rightful, sacred landscape." (Warren 1984, p. 77-81 in Bender 2009, p. 141)

Strawberry Island is in the middle of a lake, Lac du Flambeau, in northern Wisconsin and is considered to be a particularly sacred spot for the Chippewa. It is seen as the earthly dwelling place of Michabo, or Kitchi-Manitou, the Creator. (Bender 2009, p. 139, 140) Strawberries were thought to have supernatural power because they were the "first fruit," (Densmore 1929, p. 18 in Bender 2009, p. 141) "offered as food to the spirits after death" (Densmore 1929, p. 75 in Bender 2009, p. 141) and at girls' puberty ceremonies, therefore "symbolic of birth, rebirth and life." (Densmore 1929, 71-72 in Bender 2009, p. 142) In Chippewa sacred legend a Great Strawberry appears as "the 'Source' or tempter" along the 'path of the dead'" (Dewdney 1975, p. 103, 104 in Bender 2009, p. 142), "[an] obstacle...on the roadside like a huge rock." (Heming 1896, p. 154 quoted in Bender 2009, p. 142)

The meeting of Mother Earth and Father Sky (as the sun) is considered to be a sacred compact. (Bender 2004, p. 6) This is particularly true where power rests, as the sun does at a spot on the horizon during the solstices (solstice means "sun stand still"). (Grim 1983, p. 5, 6 in Bender 2009, p. 142) Solstices are seen as sacred covenants uniting "the Creator with the land and the people." (Bender 2009, p. 142)

The full moon, by definition, rises directly opposite the sunset. At the summer solstice sunset, what the Chippewa call the "strawberry moon" rises over Strawberry Island when viewed from the knoll above the cemetery and village northwest of Strawberry Island. Sacred scrolls seem to indicate the "road to paradise" or the "path of the dead" have the same orientation as the summer solstice sunset—Strawberry moonrise line. (Bender 2009 p. 142, 143)

Bender discovered that a map of "Ojibwa after life" had almost the exact location and spacing of features as a physical landscape map of Lac du Flambeau:

> Strawberry Island, Medicine Rock [a very sacred massive red granite glacial erratic located at the point of a peninsula between Strawberry Island and the village], the Bear River inlet on Lac du Flambeau and the location of the old cemetery all became earthly Map comparison and juxtaposition of the physical landscape features at Lac du Flambeau....Stretching along a line from the southeast to the northwest, Strawberry Island reflected the created earth. Medicine Rock was the "Giant Strawberry, "stand[ing] on the roadside like a huge rock"...and tempting spirits on the path to paradise. The Bear River inlet mirrored...a "deep, rapid stream..." or barrier to the soul traveling the "Ghost road" after death. (Heming 1896, p. 135 quoted in Bender p. 147)

The heavenly counterpart of the "path of the dead" or the "path of the souls" is the Milky Way, "seen by many tribes as a bridge stretching across the night sky toward the west (southwest or northwest) and along which the souls of the deceased (and shamans in full consciousness) travel (west) on their way to paradise." (Goodman 1992, p. 22-23, 38; Hadingham 1984, p. 94; Krupp 1991, p. 272-273; Russel 1980, p. 46-47; Schwartz 1997, p. 93-96 in Bender 2009, p. 147, 148) The Milky Way is also called the "serpent's path" and "the soul traveling (west) to paradise must cross" a dreaded bridge, much like a serpent, that spans the stream. (Heming 1896, p. 135 in Bender 2009, p. 148)

Participants who observed the full moon at the summer solstice from the vantage point of the knoll above the village and cemetery experienced a hierophany as the mundane was transformed into the sacred. When the moon rise reached directly over the center of Strawberry Island to the southeast, it was still low in the sky and

> its reflection created a visual path of light, representative of the Ojibwa's map of the "soul's path," mimicking on the lake how the spirit's path or road connects individual elements: creation (Strawberry Island); the Great Strawberry (Medicine Rock); the Serpent bridge (Milky Way/Bear River Inlet); and west shore of the lake near the old cemetery, or paradise on the map of the spiritual landscape. (Bender 2009, p. 151)

Figuratively, "the sky was brought down to earth"—the moonlit path connecting the significant points occurs only at that time of year because of the lower angle of the moon, "thus, once a year, transforming the landscape of Lac du Flambeau into the spiritual landscape of the 'path for souls' or 'spirits path.'" (Bender 2009, p. 152) Experiencing a hierophany with such clarity at special sites feels like the communication of sacred power and the evocation of a sacred presence. (Grim 1983, p. 5-6 in Bender 2009, p. 151, 152)

APPENDIX B

Sacred Corn

Developing a sense of the sacred in nature is one of many important challenges for ecopsychology, a process facilitated by a study of indigenous cultures. Indigenous corn cultures developed a sacred sense of nature through agriculture. The Mayans, for example, used the life cycle of maize as "the great metaphor of Maya life" and the root of their language, rituals and calendar (Fussell 1992, p. 34):

> For the Indians, accompanying step by step the corn's cycle of death and resurrection is a way of praying; and the earth, that immense temple, is their day-to-day testimony to the miracle of life being reborn. (Galeano 1987 quoted in Fussell 1992, p. 33)

All aspects of the cultivation and consumption are carefully observed and ritually celebrated in indigenous corn cultures because corn is experienced as a god and spirit upon which their survival depends. Natives will sing to their plants and calm them after earthquakes. All animals associated with corn are sacred, and the Navaho see pollen as being the essence and the Word of corn. Each phase of corn development may have a god or goddess associated with it. (Fussell 1992, p. 114, 115, 122, 328-330)

Subsistence planters created myths and rituals around the primacy of vegetative death for the generation and sustenance of new life. The parallel in human life is chosen death, self-sacrifice, to generate and sustain the social order with the belief that pain and death are inevitable and necessary. (Fussell 1992, p. 41, 42) Humans are aware of their own deaths and spiritual leaders experience spiritual transformations in association with pain, suffering and mortification of the flesh in initiation and/or sacrifice ceremonies.

Death and the new life that follows death is an archetypal theme in both Western culture and the corn cultures of the New World. Christ's death, descent into the Underworld, and resurrection and transformation get played out in a vegetative analogy in corn cultures. The phallic nature of an ear of corn and its potential generative aspects in the seeds have sexual associations. Mayans worshiped the corn ear as the face of the god, and separating the ear from the stalk and husking it were interpreted as castration, killing and sacrifice. This death of the vegetative phase was seen as simultaneously producing the spirit of corn in the ear as well as the potential for rebirth: only from the dried "dead" seed can the next generation of corn flesh/food emerge when a seed is buried (in the underworld) and moistened. This produces a new generation and a regeneration of the corn spirit as materialized spirit. (Fussell 1992, p. 30-46, 283-296) In Western alchemy, death is equally well portrayed by the death and burial of a human or by a grain being planted in the earth. Rebirth is depicted as the resurrection of a person or the sprouting of a seed. To the Mayans a single kernel of red corn represented the life of the individual, all of life in its life-and-death existence (*pars pro toto*), and the blood of sacrifice necessary to sustain and give birth to life and spirit, and the universe. (p. 31-36)

The unconscious does not distinguish between the manifest substance of animal and human flesh and the "body" of a corn plant and corn kernels. Eating the flesh of corn and the flesh of humans becomes synonymous. Ritual cannibalism is central to the indigenous planting cultures of the Western World because

> a plant rather than an animal was the supreme symbol of the life-death connection. [New World planting cultures had a] primal sense of the mutual dependence of humankind and plant-kind, the interchange of blood with sap and flesh with grain. For the same reason, these civilizations adhered to cannibalistic rituals, animal flesh could not substitute for human flesh because animals did not cultivate plants, only people did. (Fussell 1992, p. 42)

Since the energy of life is part of the energy of the universe and operates under the same systems (as seen by the indigenous mind), the change of the seasons and the cyclical movements of the stars and planets are related to human life and plant development. "As above, so below" said the alchemists. The stars and planets get associated with the spiritual realm because they are more abstract and removed.

Many corn culture natives in the New World who have been Christianized have no difficulty in simultaneously worshiping Jesus while retaining their beliefs and ceremonies associated with the corn spirit. The Greek myth of Persephone's return above ground after her annual stint in the underworld is associated with the growth of crops and Christ rising from the dead, showing the eternal nature of the spirit and of manifesting the spirit in reality. Indigenous psyches unconsciously recognize this archetypal theme while Christianity separates us from nature when it demonizes the nature spirits as being outside "The True Faith."

For a more comprehensive description of the Mayan corn cult and stories from other indigenous corn cultures, see Betty Fussell's excellent book, *The Story of Corn*, p. 29-58, 99-132 and 282-300.

APPENDIX C

Praying Mantis as a Spirit Animal

Having a spirit animal is one way of experiencing the sacred in nature as illustrated by Jungian analyst Julia Jewett's connection with the Praying Mantis. She was moved when she heard the writer Laurens van der Post describe the African Bushman's devotion to this insect. To them, the mantis is not "like" or "as if" it were a god; there is no reflective philosophical theology about it—to see a Mantis is simply to be in the presence of God. Subsequent to hearing van der Post the Praying Mantis appeared at crucial "soul junctures" in Julia's life as "testimony to that deep inner direction showing itself in the outer world, confirming my choices, validating the rightness of my path." Julia noted that "Jung called these outer manifestations of deep inner reality "synchronicity," I experience it as a profound grace." (Jewett 1966, p. 33)

Two years after hearing van der Post she had moved into a new home after her family had been raised. She had turned her attention to professional training to become a pastoral counselor. Two months after moving into her new home, she picked up what she thought was a discarded cigar in her front yard, only to discover it was a big Praying Mantis! Two years after what felt like this mantis-blessed house experience, she was completing her training program in counseling and was contemplating applying for a fellowship to stay on for a third year. Such training would have extended a program that limited the psyche to ego-consciousness, forcing her to further ignore her experience of Jung's concept of the ego relative to the greater self, "that purposeful intelligence that runs our being from an invisible Center." Though this ego-conscious focus was harsh on Julia's soul, completing a third year would make her feel more competent and would look good on her resume. While contemplating a fellowship as she drove home, she was moving toward the left turn lane on a six-lane divided highway when

> a flash of lime-spring-green caught my peripheral vision. Looking more directly at the cement divider rising in the

> middle of this concrete road, I saw Mantis. In shorter time than it takes to tell it, I knew he had about as much business being in the midst of this traffic as I did considering staying on in a soulless training venue, and I feared he would die. Suddenly I knew I would die, as well, if I continued to show up in the wrong place. (Jewett 1966, p. 33)

Her unsuccessful attempt to capture and save the mantis—it was run over by the time she got back with a bottle—left her angry and crying with grief and despair, "but chiefly [with] gratitude. And astonishment. I was dumbfounded to discover a soul friend who gave his life so that I could see my own." (Jewett 1966, p. 34) Similar experiences over a period of years led Julie to "trust completely where Mantis leads me... Paying attention to one's own nature may be the best way to show respect and gratitude to a soul friend." (Jewett 1966, p. 35)

APPENDIX D

Pink Floyd and the Fly in Life's Ointment

As an entomologist I was intrigued when I heard "Grantchester Meadows" on Pink Floyd's "Ummagumma" album (Roger Waters composer, EMI Records, ASCAP 1969). It could be interpreted through one archetypal lens as a song about the extermination complex *vis a vis* the shadow in the Western psyche. One hears a fly buzzing well into the beginning of the song. The melodic music and the words present a surface level of serene beauty:

In the lazy water meadow I lay me down.
All around me golden sun flakes settle on the ground.
Basking in the sunshine of a by-gone afternoon
Bringing sounds of yesterday into this city room.

But a hint of foreboding is present:

"Icy wind of night be gone this is not your domain"
In the sky a bird was heard to cry.
Misty morning whisperings and gentle stirring sounds
Belied the deadly silence that lay all around.

There are allusions to the underworld/unconscious as a dog fox goes underground and a kingfisher dives underwater. Consciousness is not on solid ground if one lays down in a "water meadow." A hint of insanity surfaces with the crazed laughter barely audible in the background near the end of the song. As the pleasant music continues after the last words, the fly reappears with a loud, disturbing, irritating buzz. One hears the heavy thump-thump-thump of a person descending stairs followed by a determined march towards the fly, fly swatter thrashing about. Finally a loud "swat" as the fly is zapped in what feels like one's midbrain. The pleasant bird song that had circled like a leitmotif throughout "Grantchester Meadows" now sounds like a wounded bird. The pest/symptoms have now been exterminated—or not?!

The music turns frantic, crazed, weird and driven as it seamlessly moves into the next cut on the album, "Several Species of Small Furry Animals Gathered Together in a Cave and Grooving With a Pict" (composed by Roger Waters, EMI records, ASCAP 1969). A cacophony of strange sounds surge and recede, only to mount another attack on the auditory system. Some sounds are like animals on acid, others like a ghostly human wailing. Finally a male voice utters incomprehensible words. The darker side of the unconscious has sprung forth out of the shadows with a vengeance, refusing to be exterminated by willful human intent or subdued by surface pleasantries.

Pink Floyd captured the zeitgeist of the time—1969. The Pleasantville cultural landscape of the 1950s had morphed into the American shadow on display as the whole world watched us trying to exterminate the problem in Vietnam. The victims of racial oppression in America would no longer let themselves be held down, violently asserting their power through the Black Panthers. American puritanical sexual repression was challenged with hippy free love and assaulted by the pied pipers of cock rock, the Rolling Stones. "Power to the people" was confronting the entrenched political machines, and women and Native Americans were making themselves heard in bold new ways. Demographics were dominated by adolescents and "twenty-somethings" as the Baby Boomers were coming of age, pumping adolescent uncertainty, confusion and intensity into the cultural milieu.

Cultures are made up of individuals, and the archetypal themes and issues of the conscious relationship to highly charged, un-repressable, and frightening unconscious contents found expression for many of us in the music of Pink Floyd. Although American culture as a whole displayed these themes writ large in the 60s, every individual in every culture experiences the problems of the conscious relationship to difficult unconscious material at crucial points in their lives, such as adolescence, mid-life crises, serious illness, tragedy in the family, divorce, etc. The fly cannot be eradicated from life's ointment.

REFERENCES

Achad, F. 1973. *The Egyptian Revival.* Samuel Weiser: NY.

Artemidorus. 1975. *Oneirocritica.* Robert J. White, trans. Noyes Press: Park Ridge, NJ.

Bender, H. 2004. Star Beings and Stones: Origins and Legends. In *Indian Stories Indian Histories.* Fedora Giordano and Enrico Comba, eds. Otto Editore: Torino, Italy. p. 7-22.

—— 2009. Caring for Creation: A Hierophany at Strawberry Island. In *The Archaeology of People and Territory.* George Nash and Dragos Gheorghiu, eds. Archeolingua Alapitvany: Budapest, Hungary. p. 137-156.

Berry, P. 1975. "The rape of Demeter/Persephone and neurosis." *Spring.* p. 186-198.

Beston, H. 1981. *The Outermost House.* Penguin: Harmondsworth.

Borradaile, L., Potts, F., Eastham, L. and Saunders, J. 1961. *The Invertebrata.* 4th ed. Revised by G. Kerkut. Cambridge University Press: Cambridge.

Brown, L. R. 2008. *Plan B 3.0: Mobilizing to Save Civilization.* W. W. Norton & Company: NY.

Brugsch, H. 1885. *Religion und Mythologie der alten Agypter.* Leipzig.

Cheng, F. 1994. *Empty and Full—The Language of Chinese Painting.* Shambala: Boston.

Comstock, J. 1950. *An Introduction to Entomology.* Comstock Publishing Company, Inc.: Ithaca, NY.

Densmore, F. 1918. Teton Sioux Music. *Smithsonian Institution Bureau of American Ethnology Bulletin.* Government Printing Office: Washington, DC.

—— 1929. Chippewa Customs. *Smithsonian Institution Bureau of American Ethnology Bulletin 86.* Government Printing Office: Washington, DC.

Dewdney, S. 1975. *The Sacred Scrolls of the Southern Ojibway.* University of Toronto Press: Toronto and Buffalo, NY.

Eisner, T. and Wilson, E. 1977. General Introduction: the Conquerors of the Land. In *The Insects: Readings from Scientific American.* Selected and introduced by Thomas Eisner and E. O. Wilson. W. H. Freeman and Co.: San Francisco. p. 3-15.

Elzinga, R. 2000. *Fundamentals of Entomology*. 5th ed. Prentice Hall: Upper Saddle River, NJ.

Frankfort, H. 1961. *Ancient Egyptian Religion*. Harper Torchbook: NY.

Fussell, B. 1992. *The Story of Corn*. Alfred A. Knopf: NY.

Galeano, E. 1987. *Memory of Fire II: Masks and Faces*. C. Belfrage, trans. Pantheon Books: NY.

Geertz, C. 1971. Deep Play: Notes on the Balinese Cockfight. In *Myth, Symbol, and Culture*. Clifford Geertz, ed. W. W. Norton: NY. p. 412-453.

Gimbutas, M. 1989. *The Language of the Goddess*. Harper: San Francisco.

Goethe, J. W. 1980. The Holy Longing. In *News of the Universe: Poems of Twofold Consciousness*. Robert Bly, trans. and ed. Sierra Club Books: San Francisco.

Goleman, D. 1995. *Emotional Intelligence*. Bantam Books: NY.

Goodman, R. 1992. *Lakota Star Knowledge*. Sinte Gleska College: Rosebud, SD.

Grim, J. 1983. *The Shaman: Patterns of Religious Healing Among the Ojibway Indians*. University of Oklahoma Press: Norman.

Hadingham, E. 1984. *Early Man and the Cosmos*. Walker & Co: NY.

Hemming, H. 1896. *The History of the Catholic Church in Wisconsin*. Catholic Historical Publishing Co.: Milwaukee, WI.

Hillman, J. 1982. The Animal Kingdom in the Human Dream. In *Eranos Yearbook* 1982, Vol. 51. Rudolf Ritsema, ed. Insel Verlag: Frankfurt am Main, Germany. p. 279-334.

—— 1983. *Archetypal Psychology: A Brief Account*. Spring Publications: Dallas.

—— 1988. "Going bugs." *Spring* 1988: 40-72.

—— 1989. "Cosmology for soul: from universe to cosmos." *Sphinx* 2: 17-33.

—— 1992. *The Thought of the Heart and the Soul of the World*. Spring Publications: Woodstock, Conn.

Jewett, J. 1966. "A thank you to the Praying Mantis." *Transformation: The Bulletin of the C. G. Jung Institute of Chicago* Winter 1966. p. 23-35.

Jung, C. *The Collected Works of C. G. Jung*. [CW] 2nd ed. H. Read, M. Fordham, G. Adler and W. McGuire, eds. R.F.C. Hull, trans. Princeton University Press: Princeton.

—— CW 5. 1956. *Symbols of Transformation.*

—— CW 8. 1969. *The Structure and Dynamics of the Psyche.*

—— 1961a. *Memories, Dreams, Reflections.* Aniela Jaffe, ed. Richard and Claire Winston, trans. Random House: NY.

—— 1961b. Interview. *Good Housekeeping Magazine.* December, 1961.

—— 1976. *The Visions Seminars.* From the complete notes of Mary Foote, Books One and Two. Spring Publications: Zurich.

—— 1988. *Nietzsche's* Zarathustra: *Notes of the Seminar Given in 1934-1939 by C. G. Jung.* James Jarrett, ed. Princeton University Press: Princeton.

Klausnitzer, B. 1987. *Insects—Their Biology and Cultural History.* Universe Books: NY.

Kohl, J. 1860. *Kitchi-Gami: Wanderings Round Lake Superior.* Chapman & Hall: London.

Krupp, E. 1991. *Beyond the Blue Horizon.* Oxford University Press: New York & Oxford.

Lauch, J. 2002. *The Voice of the Infinite in the Small: Re-Visioning the Insect-Human Connection.* Shambala: Boston and London.

Martin, H. 1996. Running Wolf: Vision Quest and the Inner Life of the Middle-School Student. In *Crossroads—The Quest for Contemporary Rites of Passage.* Louise Mahdi, Nancy Christopher, and Michael Meade, eds. Open Court: Chicago and LaSalle, Ill. p. 313-319.

Meier, C., ed. 2001. *Atom and Archetype: the Pauli/Jung Letters 1932-1958.* D. Roscoe, trans. Princeton: Princeton University Press.

Merritt, D. L. 2008. Sacred Landscapes, Sacred Seasons: A Jungian Ecopsychological Perspective. In *The Archaeology of Semiotics and the Social Order of Things.* George Nash and George Children, eds. BAR International Series 1833. Archaeopress: Oxford. p. 153-170.

Neumann, E. 1963. *The Great Mother.* Ralph Manheim, trans. Bollingen Series XLVII: Princeton University Press.

—— 1970. *The Origins and History of Consciousness.* R. F. C. Hull, trans. Bollingen Series XLII: Princeton University Press.

Nilsson, M. 1957. *Griechische Feste von religioser Bedeutung.* Reprint 1906, Leipzig edition. Teubner: Stuttgart.

Portmann, A. 1982. "What living form means to us." *Spring.* p. 27-38.

Rath, S. 1987. *About Cows.* Northword Press: Minoqua, WI.

Redgrove, P. 1987. *The Black Goddess and the Unseen Real*. Grove Press: NY.

Reichard, G. 1950. *Navaho Religion*. Bollingen Series II. Pantheon: NY.

Ritsema, R. and Karcher, S. trans. 1995. *I Ching*. Barnes and Noble: NY.

Russel, H. 1980. *Indian New England Before the Mayflower*. University Press of New England: Hanover, NH.

Sams, J. 1990. *Sacred Path Cards—The Discovery of Self Through Native Teachings*. Harper: San Francisco.

Sams, J. and Carson, D. 1988. *Medicine Cards—The Discovery of Powers Through the Ways of Animals*. Bear and Co.: Santa Fe.

Santino, J. 1994. *All Around the Year—Holidays and Celebrations in American Life*. University of Illinois Press: Urbana and Chicago.

Schwartz, M. 1997. *A History of Dogs in the Early Americas*. Yale University Press: New Haven & London.

Shortridge, J. 1989. *The Middle West*. University of Kansas Press: Lawrence, KS.

Southwood, T. 1997. "Entomology and mankind." *Proc. XV Internat. Cong. Entomol.* p. 35-51.

Storm, H. 1972. *Seven Arrows*. Ballentine Books: NY.

Sullivan, W. 1984. *Landprints: On the Magnificent American Landscape*. Times Books: NY.

Turner, V. 1967. *The Forest of Symbols: Aspects of Ndembu Ritual*. Cornell University Press: Ithaca, NY.

Walker, B. 1983. *The Woman's Encyclopedia of Myths and Secrets*. Harper and Row: San Francisco.

Warren, W. 1984. *History of the Ojibway People*. Minnesota Historical Society Press: St. Paul.

Wilhelm, R. 1967. *The I Ching or Book of Changes*. Cary Baynes, trans. Princeton University Press: Princeton, NY.

INDEX

D

E

F

G

H

I

J

K

L

M

N

ABOUT THE AUTHOR

Dennis Merritt, Ph.D., is a Jungian psychoanalyst and ecopsychologist in private practice in Madison and Milwaukee, Wisconsin. Dr. Merritt is a diplomate of the C.G. Jung Institute, Zurich and also holds the following degrees: M.A. Humanistic Psychology-Clinical, Sonoma State University, California, Ph.D. Insect Pathology, University of California-Berkeley, M.S. and B.S. Entomology, University of Wisconsin-Madison. Over twenty-five years of participation in Lakota Sioux ceremonies has strongly influenced his worldview.

You might also enjoy reading these Jungian publications:

Lifting the Veil
by Jane Kamerling & Fred Gustafson
ISBN 978-1-926715-75-9

Becoming by Deldon Anne McNeely
ISBN 978-1-926715-12-4

The Creative Soul by Lawrence Staples
ISBN 978-0-9810344-4-7

Guilt with a Twist by Lawrence Staples
ISBN 978-0-9776076-4-8

Enemy, Cripple, Beggar by Erel Shalit
ISBN 978-0-9776076-7-9

The Cycle of Life by Erel Shalit
ISBN 978-1-926715-50-6

Eros and the Shattering Gaze by Ken Kimmel
ISBN 978-1-926715-49-0

Divine Madness by John R. Haule
ISBN 978-1-926715-04-9

Tantra and Erotic Trance by John R. Haule
Vol. 1 ISBN 978-0-9776076-8-6
Vol. 2 ISBN 978-0-9776076-9-3

Farming Soul by Patricia Damery
ISBN 978-1-926715-01-8

The Motherline by Naomi Ruth Lowinsky
ISBN 978-0-9810344-6-1

The Sister From Below by Naomi Ruth Lowinsky
ISBN 978-0-9810344-2-3

CPSIA information can be obtained at www.ICGtesting.com
Printed in the USA
LVOW08s1604141213

365303LV00001B/72/P